This book summarizes the present state of knowledge on the microscopic behaviour of oxide surfaces.

The first chapter of the book summarizes classical approaches, introduces the concept of ionicity, and describes the mixed iono–covalent character of the oxygen cation bond in bulk materials. The next three chapters focus on the characteristics of the atomic structure (relaxation, rumpling and reconstruction effects), the electronic structure (band width, gap width, etc.) and the excitations of clean surfaces. Metal–oxide interfaces are considered in the fifth chapter with special emphasis on the microscopic interfacial interactions responsible for adhesion. The last chapter develops the concepts underlying acid–base reactions on oxide surfaces, which are used in catalysis, in adhesion science, and in colloid physics, and discusses their applicability to the adsorption of hydroxyl groups. A comprehensive list of references is included.

This book will be of interest to graduate students or researchers, to experimentalists and theorists in solid state physics, chemistry and physical chemistry.

T0183686

Physics and Chemistry at Oxide Surfaces

PHYSICS AND CHEMISTRY
AT OXIDE SURFACES

Claudine Noguera

Laboratoire de Physique des Solides, Orsay, FRANCE

CAMBRIDGE
UNIVERSITY PRESS

CAMBRIDGE UNIVERSITY PRESS
Cambridge, New York, Melbourne, Madrid, Cape Town, Singapore, São Paulo

Cambridge University Press
The Edinburgh Building, Cambridge CB2 2RU, UK

Published in the United States of America by Cambridge University Press, New York

www.cambridge.org
Information on this title: www.cambridge.org/9780521472142

First published 1996
This digitally printed first paperback version 2005

A catalogue record for this publication is available from the British Library

Library of Congress Cataloguing in Publication data

Noguera, Claudine.
Physics and chemistry at oxide surfaces / Claudine Noguera.
p.   cm.
Includes bibliographical references (p.   –   ) and index.
ISBN 0 521 47214 8 (hardcover)
1. Surface chemistry.   2. Metallic oxides–Surfaces.   I. Title.
QD506.N57   1996
546′.72153–dc20   95-48211 CIP

ISBN-13 978-0-521-47214-2 hardback
ISBN-10 0-521-47214-8 hardback

ISBN-13 978-0-521-01857-9 paperback
ISBN-10 0-521-01857-9 paperback

Nuestras horas son minutos
Cuando esperamos saber,
Y siglos cuando sabemos
Lo que se puede apprender.

Antonio Machado
Proverbios y Cantares
(Campos de Castilla)

Hours are minutes
When we hope to know,
And centuries when we know
What can be learned.

Contents

Preface

Although oxides have been the subject of active research for many years, they have attracted an increasing interest in the last decade. One reason for this interest is the discovery of superconductivity in copper oxide based materials, in 1986, with critical temperatures higher than the temperature of liquid nitrogen. Simple oxides have also been more thoroughly studied and a detailed analysis of their surface properties has been undertaken, thanks to several technological advances made during this period. The success in compensating charging effects, for example, has allowed spectroscopic measurements to be performed. Photoemission, x-ray absorption, Auger spectroscopy and low-energy electron diffraction now yield quantitative information, as they do for semi-conductors and metals. Topographic images of insulating surfaces can be recorded with an atomic force microscope. On the theoretical side, advanced numerical codes have been developed, which solve the electronic structure, optimize the geometry, and start accounting for dynamical effects in an *ab initio* way. The results presently available allow a first synthesis of the field.

The interest and the richness of the field of oxide surfaces lies in its inter-disciplinary nature and in the diversity of questions it raises, both on a fundamental and on an applied level. For example, geophysicists and geologists consider in detail the surface properties and porosity of the rocks of our earth, made of complex oxides whose properties are, to a large extent, controlled by the grain boundaries and internal surfaces. The mineral reactivity also interests the toxicologists, who try to understand the interactions between small particles and the biological medium. Small oxide particles may be found in colloidal suspensions, and the surface reactions, which take place at the interface with the solution, determine in a large part the nature and strength of the particle–particle interactions and thus the stability of the suspension. Oxides are also good catalysts, which are largely used, for example, in petrochemistry. Specific experimental techniques have been developed to quantify the acidity or basicity of their surface sites and to establish a correlation between their catalytic activity and their surface geometry. In the multi-layered materials produced in modern industry, oxides are often used as supports for thin metallic films or grafted polymeric layers, as intercalation layers in electronic devices, etc. They are often present, although in an uncontrolled way, whenever a material is in contact with the ambient atmosphere. They play a fundamental role in corrosion processes and the surface properties of real materials, such as the friction coefficient, etc., are *in fine* determined by them. Finally, they offer a large field of investigations to surface physicists, who have developed concepts and obtained many results on metals and semi-conductor surfaces, but who are still trying to answer elementary questions on oxide surface characteristics – e.g. density of states, gap width, stoichiometry of the outer layers, reconstructions, defects.

In this vast context, the present book is restricted to well-defined themes, which have been the subject of recent developments and for which a microscopic understanding is emerging. The oxides which will be discussed are 'simple' insulating oxides, with cations from the four first columns of the periodic table or from one of the transition series, provided that they are in a closed-shell configuration – such as TiO_2. Those oxides whose insulating character results from strong electron correlations will be discarded. Their *bulk* electronic structure remains a challenge for the theoreticians and little information on their surfaces is known. For a long time, all 'simple' insulating oxides were believed to be highly ionic, and were described by classical electrostatic models. More recently, electronic structure calculations have proved that most oxides are only partially ionic, but the relative weight of covalent and electrostatic interactions in these materials remains a subject of controversy. The first chapter of this book summarizes the classical approaches, makes a presentation of the concept of ionicity and proposes a quantum model, which accounts for the mixed iono–covalent character of the oxygen–cation bonds in the bulk. The three following chapters focus on the atomic structure – relaxation, rumpling and reconstruction effects; the electronic structure – band width, gap width, etc.; and the excitations of clean surfaces. Interfaces between a metal and an oxide are considered in the fifth chapter, with a special stress put on the microscopic interfacial interactions responsible for adhesion. Although the relevance of some parameters to wetting and growth processes is indicated, all the kinetic aspects – growth of a metallic deposit on an oxide or superficial oxidation of a metal – are excluded. The last chapter develops the concepts underlying acid–base reactions, which are used in catalysis, in adhesion science and in colloid physics. It points out their relevance to chemisorption processes, and more specifically to the adsorption of hydroxyl groups on oxide surfaces.

The aim of this book is to draw a panorama of the field, to ask questions, to indicate tracks for future research, rather than to present a final picture, which has still to be built up. Due to the connections with many disciplinary fields, it has not been possible to make an exhaustive bibliography. Many review papers or books are quoted, in which the reader will be able to trace the original papers. For the sake of synthesis and pedagogy, systematic features are stressed, relevant parameters are derived and a comparison is made between the concepts used in 'this' or 'that' research area. Theoretical models are proposed to synthesize the experimental or numerical results obtained on given systems and to obtain a consistent picture. Although they can only account for the gross features and the trends, they do represent an advance on the long way of the conceptualization of knowledge. The analytical support is highly inhomogeneous: some questions are developed at length; others are sketched only quickly and will leave the reader unsatisfied. It is the sign that much work remains to be done.

The French version of this book was written after a series of lectures given in 1992–3, in Orsay, to undergraduate students in solid state physics. The audience contained students preparing their theses, as well as researchers: solid state physicists, geophysicists, electrochemists and chemists, experimentalists or theoreticians. A special effort was thus made to settle the background necessary to a complete understanding, without assuming any advanced knowledge. I am deeply indebted to all the participants who encouraged me with their attendance and helped me with their questions. Many thanks are also due to my own PhD students who have shared my queries during the last years.

The final version of this book benefited greatly from discussions, comments and encouragements from some of my colleagues. Among them, I am especially grateful to J. Friedel, J. Jupille, S. and A. Chopin, P. Dubost, E. Ilisca, J. P. Pouget, J. P. Roux, A. Auroux and J. Schultz, who read part or the whole of the French draft, D. W. Jepsen who kindly read and corrected the English version, and C. Godrèche and P. Manneville whose help was invaluable in disentangling the subtleties of TEX formatting.

C. Noguera
Orsay, France

1

Introduction

This first chapter summarizes the main bulk characteristics of insulating oxides, as a prerequisite to the study of surfaces. The foundations of the classical models of cohesion are first recapitulated, and the distinction between charge-transfer oxides and correlated oxides is subsequently established. Restricting ourselves to the first family, which is the subject of this book, we analyse the mixed iono–covalent character of the anion–cation bonding and the peculiarities of the bulk electronic structure. This presentation will allow us to introduce various theoretical and experimental methods – for example, the most common techniques of band structure calculation – as well as some models – the partial charge model, the alternating lattice model – which will be used in the following chapters.

1.1 Classical models of cohesion

Ionic solids are made up of positively and negatively charged ions – the cations and the anions, respectively. The classical models postulate that the outer electronic shells of these ions are either completely filled or empty, so that the charges have integer values: e.g. O^{--} ($2p^6$ configuration) or Mg^{++} ($3s^0$ configuration). The strongest cohesion is obtained when anions and cations are piled up in an alternating way – the anions surrounded by cations and vice versa – , a stacking which minimizes the repulsion between charges of the same sign.

The hard-sphere model

In the first models, due to Born and Madelung, the ions are described as hard spheres, put together in the most compact way (Kittel, 1990).

Ionic radii The radii of these spheres are estimated from inter-atomic equilibrium distances measured in the bulk compounds. Typical values of

1

the ionic radii are: $r(O^{--}) = 1.4$ Å, $r(Mg^{++}) = 0.78$ Å. They obey some simple rules (van Meerssche and Feneau-Dupont, 1977):

- In a given column of the periodic table, the ionic radii grow with the atomic number: e.g. $r(Li^+) = 0.78$ Å $< r(Na^+) = 0.98$ Å $< r(K^+) = 1.33$ Å.
- For a given element, the ionic radii grow with the electron number: e.g. $r(Fe^{+++}) = 0.67$ Å $< r(Fe^{++}) = 0.82$ Å. This increase is associated with the expansion of the electronic shells induced by electron–electron repulsions.
- Iso-electronic ions have ionic radii which decrease as the atomic number grows, because the electrons are more localized by the attractive field of the nucleus: e.g. $r(O^{--}) = 1.46$ Å $> r(F^-) = 1.33$ Å $> r(Na^+) = 0.98$ Å $> r(Mg^{++}) = 0.78$ Å $> r(Al^{+++}) = 0.57$ Å. In a series of iso-electronic ions, anions are thus larger than cations.

The determinations of ionic radii have been constantly refined. The most recent compilations give values, not only as a function of the charge state of the atom but also as a function of its coordination number Z (Shannon, 1976). An increase in Z is always associated with an increase of the ionic radius: e.g. $r = 0.99$ Å if $Z = 4$, $r = 1.12$ Å if $Z = 7$, $r = 1.24$ Å if $Z = 9$ and $r = 1.39$ Å if $Z = 12$, for Na^+.

Madelung energies   The lattice electrostatic energy involves a sum of elementary coulomb interactions between pairs of ions (ij), bearing charges Q_i and Q_j, at a distance R_{ij}:

$$E_M = \frac{1}{2} \sum_{i \neq j} \frac{Q_i Q_j}{R_{ij}} . \qquad (1.1.1)$$

This is called the Madelung energy. For a binary crystal containing N formula units, in which the charges are equal to $-Q$ and $+nQ$ (e.g. TiO_2: $Q = 2$, $n = 2$), E_M may be written:

$$E_M = -\frac{NQ^2 \alpha}{R} , \qquad (1.1.2)$$

as a function of the smallest inter-atomic distance R. The geometric dimensionless constant α, called the Madelung constant, depends only upon the lattice type. Special care must be taken to sum up the alternating series in α. In real space, it is necessary to divide the lattice into neutral entities, without dipolar or quadrupolar moments, according to prescriptions given by Ewald (1921) and Evjen and Frank (Evjen, 1932; Frank, 1950). Summations in reciprocal space may also be performed to reach convergence more rapidly. Typical values of α are:

NaCl lattice	$\alpha = 1.747565$
CsCl lattice	$\alpha = 1.762675$
ZnS blende lattice	$\alpha = 1.6381$
rutile lattice	$\alpha = 4.816$
β-quartz lattice	$\alpha = 4.439$.

The Madelung energy per formula unit depends upon the lattice type through α, upon the square of the ionic charge and upon the first neighbour inter-atomic distance. For example, E_M/N is equal to 9 eV for NaCl, 48 eV for MgO, and 146 eV for TiO_2 rutile.

Relative stability of crystal lattices   The hard-sphere model introduced by Born and Madelung was later used to understand the relative stability of crystal structures. Some structures, typical of binary oxides, are represented in Fig. 1.1. An important parameter is the ratio r_+/r_- between the cation and the anion ionic radii (r_+ and r_-, respectively). When cations

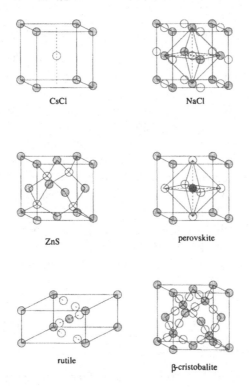

Fig. 1.1   Some binary oxide structures.

are large enough to prevent anion–anion contact, the first neighbour inter-atomic distance R is fixed by the smallest anion–cation distance and reads $R = r_+ + r_-$. If this is not the case, cations may move in cavities whose size depends only upon the anion radius; then R is fixed by r_-. This regime occurs when r_+/r_- becomes smaller than a critical ratio $(r_+/r_-)_c$, a function of the lattice type. For example, $(r_+/r_-)_c$ is equal to 0.73, 0.41 and 0.22 respectively for the CsCl, NaCl and ZnS structures. The Madelung energy E_M is equal to:

$$E_M = -\frac{NQ^2\alpha}{r_-(1 + r_+/r_-)}, \tag{1.1.3}$$

when r_+/r_- is larger than $(r_+/r_-)_c$, and to:

$$E_M = -\frac{NQ^2\alpha}{r_-\left[1 + (r_+/r_-)_c\right]}, \tag{1.1.4}$$

otherwise. The variations of E_M as a function of r_+/r_- are represented in Fig. 1.2 for three simple cubic structures: CsCl (coordination number $Z = 8$), NaCl ($Z = 6$) and ZnS ($Z = 4$). The intersection points of the three curves define the limits of stability for each structure:

ZnS structure	$r_+/r_- < 0.35$
NaCl structure	$0.35 < r_+/r_- < 0.70$
CsCl structure	$0.70 < r_+/r_-.$

Structures with increasing coordination numbers Z correspond to increasing r_+/r_- ratios. The law is qualitatively well obeyed along series of compounds which involve the same anion: for example, in the series LiCl, NaCl, KCl, RbCl and CsCl, the first four compounds display a NaCl structure, while the last one crystallizes in a CsCl structure. In the series NaI, AgI, CsI, along which r_+ increases, a ZnS structure is found for AgI ($r_+/r_- = 0.51$), a NaCl one for NaI ($r_+/r_- = 0.45$) and a CsCl one for CsI ($r_+/r_- = 0.75$). Conclusions qualitatively similar apply to compounds of stoichiometry MX_2 which crystallize in structures such as α-quartz (SiO_2: $Z = 4$), rutile (TiO_2: $Z = 6$) or fluorine (CaF_2: $Z = 8$):

α-quartz structure	$r_+/r_- < 0.41$
rutile structure	$0.41 < r_+/r_- < 0.73$
fluorine structure	$0.73 < r_+/r_-.$

A discussion of lamellar structures and ternary systems may be found in van Meerssche and Feneau-Dupont (1977).

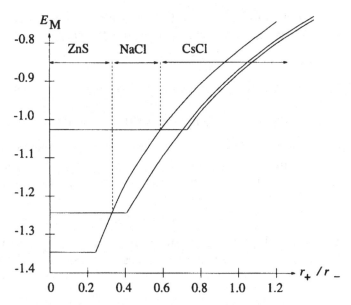

Fig. 1.2. Madelung energy for the three cubic structures ZnS, NaCl and CsCl as a function of the ratio r_+/r_- between the anion and cation ionic radii, at fixed values of r_-. The quantity $-\alpha/(1+r_+/r_-)$, proportional to E_M, is shown along the vertical axis. The intersection points of the three curves define the limits of stability for each structure.

The iodine series, which was quoted above, proves that the model works well, except in the vicinity of the critical values, where some discrepancies may appear. This shows that the non-coulombic contributions to the cohesive energy, although small in absolute value, may become relevant when tiny energy differences exist. We will discuss the origin of these interactions in Section 1.2.

Pauling's rules   The preceding discussion provides an understanding of the four empirical rules which, according to Pauling, control the electrostatic stability of ionic compounds.

- Cations are surrounded by an anion polyhedron. The shortest distance between an anion and a cation is equal to the sum of the ionic radii. The ratio r_+/r_- fixes the coordination.
- The sum of the anion–cation bond strengths, around a given anion, is equal to the charge of this ion. This rule allows a prediction of the anion coordination number in complex structures, such as those met in ternary compounds. It makes use of the concept of bond strength, which, according to Pauling's definition, is equal to the cation charge divided by its coordination number.

- Coordination polyhedra are generally linked by vertices. This is especially true when the cation charge is high and the cation coordination number low. When edges or faces are shared, the structure stability decreases.
- In a crystal with several types of cations, the polyhedra around those which have the higher charge or the lower coordination, tend to avoid each other. This rule is well exemplified in the perovskite structures.

This discussion stresses that, as a first approximation, cohesion is governed by coulomb interactions and that the prediction of equilibrium structures only requires a knowledge of the charges and ionic radii.

Born's model and later developments

The hard-sphere model, presented above in its most simplified form, was later refined in order to account for the cohesion energy and the elastic properties of ionic crystals.

Short-range repulsion A first improvement is related to the repulsive forces which become effective at short inter-ionic distances. In the original model, steric – or hard-core – repulsive forces prevent two ions i and j from coming closer than the sum of their ionic radii r_i and r_j. The short-range repulsion energy is infinite if $R_{ij} < r_i + r_j$, and zero otherwise, which may be written in the form:

$$V_{ij} = \left(\frac{r_i + r_j}{R_{ij}} \right)^n , \tag{1.1.5}$$

with $n = \infty$. The existence of repulsive interactions, impeding inner electronic shells from overlapping, results from Pauli's principle. Yet, there exists no analytical expression deduced from first principles to account for it. Depending upon the fields of research, various empirical laws are used, among which the Lennard–Jones law:

$$V_{ij} = \frac{A}{R_{ij}^n} , \tag{1.1.6}$$

and the Born–Mayer one:

$$V_{ij} = B \exp \left(-\frac{R_{ij}}{\rho} \right) . \tag{1.1.7}$$

Each of these contains two parameters which are functions of the interacting ions i and j: A and n in (1.1.6), with $5 \leq n \leq 12$; B and ρ in (1.1.7), with ρ of the order of 0.2 to 0.3 Å. In both cases, the repulsion energy decreases strongly when the inter-atomic distance gets larger, so that only first neighbour interactions are relevant. The Lennard–Jones form is often used to describe the cohesion of van der Waals systems, for example rare gas crystals or clusters. In the physics of metals or insulators, one usually chooses the Born–Mayer law, which has a better microscopic basis, since exchange interactions involve atomic wave functions which exponentially decrease far from the nucleus.

Born's model, or rigid-ion model In Born's model, the total energy of a binary compound, containing N formula units, is thus equal to the sum of the Madelung energy E_M and the short-range repulsion energy between first neighbours (Born and Huang, 1954; Tosi, 1964):

$$E = -\frac{NQ^2\alpha}{R} + NZB\exp\left(-\frac{R}{\rho}\right) . \qquad (1.1.8)$$

The crystal structure determines the Madelung constant α and the number Z of anion–cation bonds per formula unit. In binary compounds, Z is equal to the coordination number of the ion which has the larger number of first neighbours (e.g. $Z = 6$ for TiO_2). The minimization of E with respect to R yields the equilibrium distance R_0 between first neighbours in a given structure. R_0 is the solution of the implicit equation:

$$ZB\exp\left(-\frac{R_0}{\rho}\right) = \frac{\rho\alpha Q^2}{R_0^2} . \qquad (1.1.9)$$

The cohesion energy, necessary to separate the system into independent ions, is equal to:

$$E_{coh} = N\frac{\alpha Q^2}{R_0}\left(1 - \frac{\rho}{R_0}\right) . \qquad (1.1.10)$$

The first factor in the parenthesis is associated with E_M and the second with the short-range interactions. Since the ratio between ρ and R_0 is generally of the order of 0.1, the largest contribution to the cohesion energy is E_M, which justifies *a posteriori* the hard-sphere model. Despite its simplicity, Born's model has been used with success in many different instances; for example, it has helped in the interpretation of many bulk phonon dispersion curves (Bilz and Kress, 1979).

Improvements Further improvements have been introduced, as experimental data have become more reliable. For example, neutron scattering experiments have proved that the longitudinal optic mode frequencies at the zone centre are systematically lower than predicted. Various models result from the idea that this discrepancy may be assigned to the neglect of ionic polarization.

- The shell model: the ions are described by a core, including the nucleus and the inner electrons, and a zero-mass shell representing the valence electrons (Dick and Overhauser, 1958). The core and the shell bear opposite charges. They are harmonically coupled by a spring of stiffness k. The electric field $\mathscr{E}$ exerted by neighbouring ions shifts the shell with respect to the core position. If Y is the shell charge, $\mathscr{E}$ induces a dipole moment equal to $\mathscr{E}Y^2/k$. The ion polarizability α' in the model, is thus equal to: $\alpha' = Y^2/k$. The value of k may be deduced from the value of the optical dielectric constant ϵ_∞ thanks to the Clausius–Mosotti relationship:

$$\frac{\epsilon_\infty - 1}{\epsilon_\infty + 2} = \frac{4\pi\alpha'}{3} . \tag{1.1.11}$$

Only short-range inter-ionic forces between the shells are taken into account, while long-range forces involve all species, except the core and shell associated with the same ion. The weak polarizability of the cations is very often neglected.

- Other refinements: in the literature, many other refinements to Born's model have been introduced: among others, one finds models which take into account the mutual ion polarizability (van der Waals interactions), the extended shell model, the overlap shell model, the deformable shell model, the breathing shell model and the double shell model (Cochran, 1971). In covalent crystals such as diamond, where the charges are not located on the sites but rather on the bonds, the so-called bond charge model was proposed. Some authors found it necessary to introduce non-integer charges in some systems: in a classical approach, this raises the question of potential transferability and of the nature of the quantum terms responsible for the charge transfers. During the last few years, an increasing effort has been put on the derivation of inter-ionic potentials from *ab initio* methods (e.g. Allan *et al.*, 1990; Harding, 1991; Harrison and Leslie, 1992; Purton *et al.*, 1993; Allan and Mackrodt, 1994). The classical models are generally recognized to be less well suited to open shell systems, like transition metal oxides, than to simple oxides (Stoneham and Harding, 1986). More details on all these models may be found in specialized papers on this subject (Bilz and Kress, 1979). In most cases, it is likely that the increasing complexity of classical models simply hides the need to treat the quantum effects correctly.

Applications of the atomistic models

The pair potentials have been widely used to describe various cohesion properties of insulating compounds:

- bulk static properties, such as the cohesion energy, the thermal expansion coefficient, the relative stability of polymorphs (Catlow and Mackrodt, 1982),
- structural phase transitions, such as the anti-ferrodisplacive cubic–tetragonal transition in SrTiO$_3$, the tetragonal–orthorhombic transition in La$_2$CuO$_4$ (Piveteau and Noguera, 1991) or the amorphization of α-quartz under pressure (Binggeli *et al.*, 1994),
- bulk dynamic properties: the phonon dispersion curves, the bulk modulus and the elastic coefficients (Bilz and Kress, 1979),
- thermodynamics of defects: the energies of formation of defects, the vacancy migration energy (Harding, 1990), the thermodynamics of non-stoichiometric oxides (Harding, 1991; Boureau and Tetot, 1989), the simulation of superionic conductors (Lindan and Gillan, 1994),
- surface static properties: the surface tension, the relaxation effects (Catlow and Mackrodt, 1982; Mackrodt, 1988),
- surface dynamic properties: the phonon frequencies, the vibrational entropy, the mean square displacements (Kress and de Wette, 1991),
- surface defects: steps, kinks, doping, surface vacancies (Mackrodt, 1984; Stoneham and Tasker, 1988; Colbourn, 1992),
- interfacial properties (Stoneham and Tasker, 1988),
- small cluster properties (Martins, 1983; Ziemann and Castelman, 1991).

1.2 Origin of the insulating state

While the classical models can reproduce and sometimes predict some structural properties, they are unable to inform about the electronic characteristics of insulators, because they assume that the electrons are frozen around the ionic cores. The next step consists in finding the microscopic origin of the forbidden gap present in the electronic excitation spectrum, which is the defining property of the insulating state.

Charge-transfer oxides; correlated oxides

The gap width is fixed by the electronic excitation of lowest energy. Starting from a classical model with localized electrons, two types of excitations may be considered.

Charge-transfer excitation   When an electron is transferred from an anion to a cation, a charge-transfer excitation is produced:

$$X^{n+} + O^{--} \rightarrow X^{(n-1)+} + O^{-} . \tag{1.2.1}$$

The charge-transfer energy Δ is related to the cation nth ionization potential I_n and to the oxygen second electronic affinity A_2. To a first approximation, Δ reads:

$$\Delta = A_2 - I_n - V_X + V_O , \qquad (1.2.2)$$

if the two ions are infinitely far from each other, or:

$$\Delta = A_2 - I_n - V_X + V_O - \frac{1}{R} , \qquad (1.2.3)$$

if they are located at a distance R; in these expressions, V_X and V_O are the electrostatic potentials acting on a cation ($V_X < 0$), and on an oxygen anion ($V_O > 0$). For MgO, in which $A_2 = -9$ eV, $I_2 = 15$ eV, $V_O = -V_X = 24$ eV and $R = 2.1$ Å, $\Delta = 17$ eV, while for NaCl, $\Delta = 11.5$ eV ($A_1 = 3.76 \pm 2$ eV, $I_1 = 5.14$ eV, $V_O = -V_X = 8.9$ eV and $R = 2.82$ Å). For NiO: $A_2 = -9$ eV, $I_2 = 18.15$ eV, $V_O = -V_X = 24.1$ eV and $R = 2.09$ Å, which yields $\Delta = 14$ eV. The values found for Δ, in this simple approximation, are larger than the measured gap widths, which are respectively equal to 7.8 eV, 8.9 eV and 4 eV for the three compounds.

Cation charge fluctuation   A second type of excitation may occur when two cations exchange an electron. Its energy U is associated with the reaction:

$$X^{n+} + X^{n+} \rightarrow X^{(n-1)+} + X^{(n+1)+} . \qquad (1.2.4)$$

It is related to the $(n + 1)$th and nth cation ionization potentials I_{n+1} and I_n, corrected by an electrostatic term; U reads, to a first approximation:

$$U = I_{n+1} - I_n - \frac{1}{R} . \qquad (1.2.5)$$

This approach yields U close to 13 eV in NiO ($I_3 = 36.16$ eV, $I_2 = 18.2$ eV and $R = 2.96$ Å), a value slightly lower than the charge-transfer excitation energy. The cation charge fluctuations have to be considered whenever the ions have an unfilled outer electronic shell. Otherwise, the energy difference $I_{n+1} - I_n$ is very large and cannot be the lowest excitation energy of the system.

In a compound, when U happens to be large, the electronic structure can no longer be described by a one-electron effective Hamiltonian, and the mean-field approaches for electron–electron interactions are not valid. In a mean-field approximation, all the electronic configurations on a given ion have equal probabilities. For example, the two-electron wave function of the hydrogen diatomic molecule involves the configurations H–H and H^+H^- with equal weight, in the bonding state. Beyond the mean-field approximation, the configuration H^+H^- has a decreasing weight as U gets larger: the electrons move in a correlated way to avoid being located on the same atom. A partial localization results, which the band structure calculations do not reproduce well (Mott, 1974). In the physics of insulating oxides, this is what distinguishes correlated oxides from simple oxides.

In the calculation of electronic structures, the presence of correlations thus always represents a difficulty. Perturbation expansions can account for the two extreme cases: the delocalized limit in which the effective repulsion U is low compared to the band width, and the quasi-atomic limit where the electron delocalization modifies only slightly the correlated ground state (Anderson, 1959). Some variational techniques (Hubbard, 1964; Gutzwiller, 1965) allow a treatment of systems with U of the order of β, but they are difficult to use. New methods have recently been developed for adding a part of the Hubbard Hamiltonian to the LDA (local density approximation) ground state (Czyzyk and Sawatzky, 1994).

Sawatzky–Zaanen's diagram and discussion

Zaanen and Sawatzky (Zaanen *et al.*, 1985) have clarified the distinction between charge-transfer oxides and correlated oxides, by studying the excitation spectrum of nickel oxide. NiO represents the historical example of a material in which the importance of correlation effects leads to a failure of effective one-electron approaches. Until very recently, even the most elaborate band structure calculations were unable to reproduce its gap width, equal to 4 eV (Powel and Spicer, 1970; Hüfner *et al.*, 1992). Other oxides present the same peculiarities: some vanadium oxides such as V_2O_3, and also La_2CuO_4, parent of the first family of high-T_c superconductors. Sawatzky and Zaanen have proposed a phase diagram for insulating systems, relying upon the values of Δ and U. This diagram is reproduced on Fig. 1.3. A diagonal line splits it into two parts and two lines, one horizontal and one vertical, are added close to the axes to account for the finite band widths W and w associated with anion–cation and cation–cation electron delocalization, respectively. Two regions are of main importance:

- If $U > \Delta$, the anion–cation charge-transfer energy Δ is the lowest excitation energy. When $\Delta > W/2$, the system is a charge-transfer insulator and its gap width $\Delta - W/2$ reflects the value of the anion electronegativity, for a given cation and a given crystallographic structure. For example in the series $NiCl_2$, $NiBr_2$, NiI and NiS, the gaps are respectively equal to 4.7 eV, 3.5 eV, 1.7 eV and 0 eV. In the limit $\Delta < W/2$, the system is metallic.

- The Mott–Hubbard regime is located in the region $U < \Delta$: compounds are insulating if $U > w$ and metallic otherwise. Along a transition metal series, the lowering of the d level is responsible for a progressive decrease

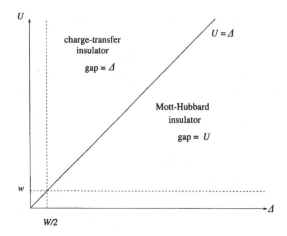

Fig. 1.3. Sawatzky–Zaanen's diagram, giving the nature of the insulating state as a function of the cation–cation charge-fluctuation energy U and the energy Δ for anion–cation charge-transfer. W and w are the band widths associated with anion–cation and cation–cation electron delocalization, respectively.

of Δ. The variations of U, on the other hand, are relatively ill known, because this parameter hides various phenomena non-directly related to the d orbitals (Friedel and Noguera, 1983). It is generally accepted that early transition metal oxides are Mott–Hubbard insulators while the late oxides are charge-transfer insulators. Resonant photo-emission may be used to check the weight of cation states at the top of the valence band. A non-negligible weight implies that the insulator is of the Mott–Hubbard type (Li *et al.*, 1992). To estimate the gap width, expressions (1.2.2), (1.2.3) and (1.2.5) are too approximate and may lead to errors both in the order of magnitude of the gap width and on the origin of the forbidden band. For example, in NiO, they give: $\Delta = 14$ eV and $U = 13$ eV. Yet it is now well established that the gap is of charge-transfer origin. The predicted trends are nevertheless correct when one compares compounds with similar structures, which differ only by the nature of a single ion (Zaanen and Sawatzky, 1990).

Nevertheless, the rule that early transition metal oxides are insulators of the Mott–Hubbard type should not be applied without care (Cox, 1992; Jollet *et al.*, 1992; Henrich and Cox, 1994). In the classical models, the cations of some early transition metal compounds have a d^0 outer electronic configuration: this is the case for titanium in TiO_2 – configuration $4s^0 3d^0$ – or for yttrium in Y_2O_3. Two charge-transfer excitations are then required to bring two cations in a d^1 state, before a charge fluctuation may occur $d^1 + d^1 \rightarrow d^0 + d^2$ with an energy U. The total energy of this process is thus equal to $2\Delta + U$; it is always much larger than Δ. The effect is similar for cations with a d^{10} configuration. Transition metal oxides with configurations d^0 or d^{10} are generally charge-transfer insulators.

In this book, we will mostly consider charge-transfer oxides whose electronic structure is well described by one-electron effective Hamiltonians and in which the contribution of correlation effects to the total energy is small.

1.3 Ionicity of a chemical bond

To correctly account for the electronic structure and the partial ionicity of an insulating oxide, classical models are of no help and the hybridization between anion and cation orbitals and the long-range electron delocalization have to be treated. Several arguments confirm this point. First, there has to exist a continuous transition, via band theory, between insulators and semi-conductors, since the border line between these two families is ill defined. On the other hand, we have mentioned above that some atomistic calculations make use of non-integer charges (Prade *et al.*, 1993). Such values may only be found if electrons are not fully localized. Similarly, in colloid science and in the field of acid–base properties of surfaces, electrostatic models which try to account for the reactivity of the surface sites as a function of their environments, have to assume that charges are non-integer and depend upon the coordination number (Chapter 6). Finally, photoemission experiments have shown that, in insulating oxides, valence band widths may reach 4 to 10 eV, which reveals the presence of important electron delocalization effects (Kowalczyk *et al.*, 1977; Tjeng *et al.*, 1990).

Nevertheless, it is not conceptually easy, either by theory or by experiment, to assign well-defined values to the ionic charges. X-ray diffraction experiments on single crystals or electronic structure calculations give electronic density maps. But the final charge values depend upon how one chooses to share the electronic density, a difficult problem especially when a bond formation implies an electron sharing between two atoms (Bader, 1991). It is thus not very rigorous to discuss the charge values, in the absolute. Yet, the concept of ionicity of the chemical bond underlies many discussions in the field of insulating oxides. We will thus recall some pioneer work: Pauling's thermo-chemical approach and Phillips' spectroscopic description.

Thermo-chemical description

Pauling (1960) proposed that the ionicity of an AC chemical bond can be characterized by a parameter f_{AC} related to the enthalpies of formation of the three diatomic molecules AA, CC and AC, i.e. to measurable thermodynamic quantities. The starting point relies upon the fact that the energy of formation E_{AC} of an hetero-polar molecule AC is larger than the mean value $(E_{AA} + E_{CC})/2$ for the homo-polar molecules, because of the electron transfer from the more electropositive to the more electronegative atom. Introducing a scaling factor γ, Pauling defines the relative electronegativities χ, by the relation:

$$(\chi_A - \chi_C)^2 = \gamma \left(E_{AC} - \frac{E_{AA} + E_{CC}}{2} \right) . \tag{1.3.1}$$

In Pauling's scale, χ is equal to 3.5 for oxygen, 1.8 for silicon, 1.5 for titanium and aluminium and 1.2, 1.0, 1.0 and 0.9 for magnesium, calcium, strontium and barium, respectively. As $\chi_A - \chi_C$ increases, the bond covalency decreases. In order to quantify the latter by a number between 0 and 1, Pauling defines the bond covalency parameter by:

$$g_{AC} = \exp \left[-\frac{(\chi_A - \chi_C)^2}{4} \right] , \tag{1.3.2}$$

and, in a symmetric way, the bond ionicity parameter by:

$$f_{AC} = 1 - \exp \left[-\frac{(\chi_A - \chi_C)^2}{4} \right] . \tag{1.3.3}$$

Typical values of f_{AC} are 0.73 for MgO, 0.79 for CaO and SrO and 0.59 for ZnO. In hetero-polar diatomic molecules, there exists a rough correlation between the values of f_{AC} and the dipole moments. Expression (1.3.3) may be generalized to solids (Pauling, 1971) to take into account the fact that each atom takes part in several bonds. If x is the cation valency and Z its coordination number in the solid, the total cation covalency, $x \exp[-(\chi_A - \chi_C)^2/4]$, is shared between the Z bonds, which yields an ionicity equal to:

$$f_{AC} = 1 - \frac{x}{Z} \exp \left[-\frac{(\chi_A - \chi_C)^2}{4} \right] . \tag{1.3.4}$$

The ionic character is thus generally larger in a solid than in a diatomic molecule, because the coordination number Z is higher in the former. For example, $f_{AC} = 0.73$ and 0.84 for the MgO molecule ($Z = 1$) and the bulk MgO ($Z = 6$), respectively. Other electronegativity scales have been proposed. In Chapter 6, we will mention Mulliken's scale (Mulliken, 1951; 1952a; 1952b) and Sanderson's scale (Sanderson, 1960), the latter being often used in the field of heterogeneous catalysis.

Spectroscopic description

Phillips (1970) criticized the thermo-chemical approach and proposed an ionicity scale founded on spectroscopic measurements. Phillips relates the energy E_g of a semi-conductor, that he calls 'gap', to two terms. The first one E_h, called the hetero-polar or ionic energy, is associated with the anion–cation difference in electronegativity. The second E_c, called the covalent energy, depends upon the electron delocalization. Thus:

$$E_g = \sqrt{E_h^2 + E_c^2} \,. \tag{1.3.5}$$

E_g is not equal to the gap width $\varDelta$, which is the smallest energy difference between filled and empty states. It is larger than $\varDelta$ and qualitatively represents the energy difference between the mean valence and conduction band positions. Its value may be deduced from the optical dielectric constant ϵ_∞ and from the plasma frequency ω_p (see Chapter 4).

Phillips characterizes the bond ionicity by the ratio f:

$$f = \frac{E_h^2}{E_h^2 + E_c^2} \,. \tag{1.3.6}$$

As in the thermo-chemical description, f is equal to 1 for ionic bonds and 0 for covalent bonds. An ionicity scale may be established based on comparative gap measurements. In the special case of an hetero-polar AC molecule, using the simplest quantum approach, we will now prove that Phillips' f parameter is equal to the square of the ionic charge.

The hetero-polar diatomic molecule (non-self-consistent treatment)

The electronic structure of a hetero-polar diatomic molecule exemplifies the concept of bond ionicity and charge transfer. In the simplest quantum approach, one accounts for the electronic structure by using a one-electron Hamiltonian H and considering a single outer orbital per site, of energy ϵ_A and ϵ_C for the anion and cation, respectively. Let us assume that these orbitals are non-degenerate and orthogonal ($\langle A|C \rangle = 0$), and let us call $\beta = \langle A|H|C \rangle$ the resonance (or hopping) integral associated with electron delocalization between neighbouring atoms. The bonding and anti-bonding states of the AC molecule are expanded in the atomic orbital basis set:

$$|\psi\rangle = x|A\rangle + y|C\rangle \,, \tag{1.3.7}$$

and their eigenenergies E are solutions of the secular equation:

$$(\epsilon_A - E)(\epsilon_C - E) - \beta^2 = 0 \,, \tag{1.3.8}$$

i.e.:

$$E_{B,AB} = \frac{\epsilon_A + \epsilon_C}{2} \pm \frac{1}{2}\sqrt{(\epsilon_C - \epsilon_A)^2 + 4\beta^2} \,. \tag{1.3.9}$$

The '−' sign is associated with the filled bonding state (B) and the '+' sign to the empty anti-bonding states (AB). The energy difference between the two molecular levels is the excitation gap E_g:

$$E_g = \sqrt{(\epsilon_C - \epsilon_A)^2 + 4\beta^2} \ . \tag{1.3.10}$$

The link with Phillips' spectroscopic model becomes apparent: $\epsilon_C - \epsilon_A$ is the hetero-polar (ionic) contribution to the gap, while 2β is the covalent contribution.

In the bonding state, the anion and cation electron numbers N_A and N_C, respectively, are equal to:

$$N_A = 2x^2 = 1 + \frac{\epsilon_C - \epsilon_A}{\sqrt{(\epsilon_C - \epsilon_A)^2 + 4\beta^2}} \ , \tag{1.3.11}$$

and:

$$N_C = 2y^2 = 1 - \frac{\epsilon_C - \epsilon_A}{\sqrt{(\epsilon_C - \epsilon_A)^2 + 4\beta^2}} \ . \tag{1.3.12}$$

If one assumes for example that both neutral A and C atoms bear one outer electron, $1 - N_A$ and $1 - N_C$ determine the ionic charges. The charge is negative on the anion and positive on the cation. Its absolute value is equal to:

$$Q = \frac{\epsilon_C - \epsilon_A}{\sqrt{(\epsilon_C - \epsilon_A)^2 + 4\beta^2}} \ . \tag{1.3.13}$$

The parameter which controls the bond ionicity is thus $(\epsilon_C - \epsilon_A)/2|\beta|$:

- When $(\epsilon_C - \epsilon_A)/2|\beta| = 0$, the charge is equal to zero. The two atoms are neutral and the bond is purely covalent: the two electrons have equal probability of being located on each atom.
- When $(\epsilon_C - \epsilon_A)/2|\beta| \to \infty$, i.e. when the hopping probability is zero or when the energy difference $\epsilon_C - \epsilon_A$ is very large, $Q = 1$. The two electrons are located on the anion: the bond has a pure ionic character.

Phillips' f parameter for the molecule is thus identical to the square of the charge: $f = Q^2$, i.e. to the charge distribution asymmetry between the anion and the cation (Garcia and Cohen, 1993).

It is interesting to consider the different contributions to the energy of the molecule. One can expand the electronic contribution in the following way:

$$E_{el} = 2E_B = 2\langle\psi_B|H|\psi_B\rangle = 2\langle\psi_B|H_D|\psi_B\rangle + 2\langle\psi_B|H_{ND}|\psi_B\rangle \ , \tag{1.3.14}$$

by splitting the Hamiltonian into a diagonal operator H_D which involves the atomic orbital energies, and a non-diagonal part H_{ND} associated with the inter-atomic resonance integrals. The two last terms in (1.3.14) are equal to:

$$2\langle\psi_B|H_D|\psi_B\rangle = 2x^2\epsilon_A + 2y^2\epsilon_C = \epsilon_A + \epsilon_C + Q(\epsilon_A - \epsilon_C), \quad (1.3.15)$$

and:

$$2\langle\psi_B|H_{ND}|\psi_B\rangle = 4xy\beta = -2|\beta|\sqrt{(1-Q)(1+Q)}. \quad (1.3.16)$$

The cohesion energy of the molecule $E_{coh} = \epsilon_C + \epsilon_A - E_{el}$, estimated with respect to the neutral atoms, thus contains two positive contributions:

$$E_{coh} = Q(\epsilon_C - \epsilon_A) + 2|\beta|\sqrt{(1-Q)(1+Q)}. \quad (1.3.17)$$

The first term, called the internal energy E_{coh}^{int} in the following, is due to the excess – for the anion – or to the loss – for the cation – of electrons on the outer atomic levels as a result of the charge transfer. The second term, E_{coh}^{cov}, is the covalent energy due to the electron delocalization. In the ionic limit, the internal energy is a maximum while the covalent term is equal to zero. In the covalent limit, it is the other way around.

This simple example proves that, whenever a charge transfer occurs, at least two terms have to be considered in the total energy: the internal energy, sometimes called the charge-transfer energy, and the covalent term. One should also add the terms due to the electron–electron interactions, absent in this non-self-consistent approach, and the short-range repulsion term. At the end of Section 1.4, a more complete approach will be presented.

1.4 Bulk electronic structure of simple oxides

The electronic structure of an hetero-polar diatomic molecule presents many features which are similar to those of an insulating crystal. When several molecules are brought together, the molecular orbitals broaden by resonance and give rise to two non-overlapping bands: the valence band and the conduction band, as indicated in Fig. 1.4. The electrons fill all the valence states while the conduction states are empty: the Fermi level lies in the forbidden energy band. In the following paragraphs, after a brief review of the experimental and theoretical techniques commonly used in the field, we will detail the electronic characteristics of insulating oxides in the bulk: the densities of states in the valence and conduction bands, the gap width and the values of the ionic charges.

Experimental techniques

Electrons of various energies, as well as photons in the x-ray or ultraviolet energy ranges, are used to probe the electronic structure. The spectroscopic

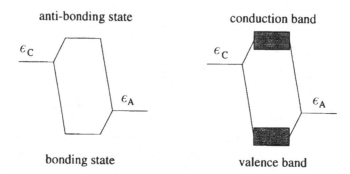

anti-bonding state   conduction band

bonding state   valence band

Fig. 1.4. Bonding and anti-bonding states of an hetero-polar diatomic molecule, in a simple two-orbital model (*left*), and broadening of these states, leading to the formation of the valence and conduction bands in the crystal (*right*).

methods are generally more difficult to implement in insulators than in metals or semi-conductors, because of the charging effects which result from the insulating character of the samples.

Electron energy loss experiments give information about the gap states, the position and shape of the deep level absorption thresholds and the collective plasmon excitations. The gap widths of simple oxides typically range from 3 to 10 eV:

NiO :	$\Delta = 4$ eV	TiO_2 :	$\Delta = 3.1$ eV
MgO :	$\Delta = 7.8$ eV	SnO_2 :	$\Delta = 3.2$ eV
CaO :	$\Delta = 7$ eV	SiO_2 :	$\Delta = 8.6$ eV
SrO :	$\Delta = 6.3$ eV	Al_2O_3 :	$\Delta = 8$ eV
BaO :	$\Delta = 4.9$ eV	Y_2O_3 :	$\Delta = 5$ eV .

When the incident electrons have large kinetic energies E_{kin}, core electrons may be excited. At small collection angles and when E_{kin} is much larger than the core level binding energy, the excitation probability for an electron of momentum l_i obeys a dipolar selection rule, $l_f = l_i \pm 1$: the threshold shape then reflects the unoccupied density of states, projected on the excited atom orbitals and on the angular momenta l_f. Ahn and Krivanek (1983) have published a compilation of absorption thresholds in many oxides. There exists a qualitative correlation between the threshold shape and the symmetry of the environment of the excited atom. In addition, the threshold energy gives information on the ionization state of the excited atom. The density of states in the conduction band may also be obtained by inverse photoemission (Taverner *et al.*, 1993).

Photon spectroscopies include photoemission in the x- and ultraviolet photon ranges (XPS and UPS, respectively) and x-ray absorption. In

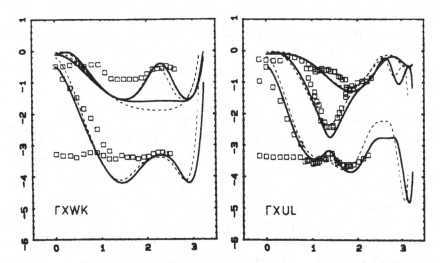

Fig. 1.5. MgO band structure, obtained by angle-resolved photoemission, along two directions in reciprocal space; *left*: ΓXWK; *right*: ΓXUL. Energies on the vertical axis are in eV and wave vectors on the horizontal axis in Å⁻¹. Full line curves are bands calculated by a tight-binding approach (after Tjeng *et al.*, 1990).

a photoemission experiment, an electron is photoemitted from a filled atomic level into vacuum, and the measured signal – number of electrons in the energy range $[E, E+dE]$ – reflects the valence band shape and/or the core level positions. In x-ray absorption, an electron is excited from an inner atomic level into an empty state above the Fermi level. The x-ray absorption cross-section may be correlated with the empty state density in the conduction band, as in electron energy loss spectroscopy. In most insulating oxides, the measured valence band width is of the order of that found in metals, from 5 to 10 eV (Kowalczyk *et al.*, 1977). Angle-resolved UPS experiments allow the band dispersion to be plotted along high-symmetry lines of reciprocal space. This has been done for MgO (Tjeng *et al.*, 1990), SrTiO$_3$ (Brookes *et al.*, 1986) and TiO$_2$ (Raikar *et al.*, 1991; Hardman *et al.*, 1994). Similar studies have been performed more systematically and for a longer period of time in metals (Thiry *et al.*, 1979). Fig. 1.5 reproduces the experimental valence band dispersion in MgO, along two directions of the reciprocal space, superimposed on a tight-binding band structure calculation. A wide band clearly shows up, which proves that the electrons in MgO have a finite effective mass and are not as localized as the classical models assume.

Theoretical techniques for electronic structure determination

Various theoretical methods have been used to determine the electronic structure of insulating oxides. We will review here only those which treat the electron–electron interactions in a self-consistent way, an essential requirement in these compounds. From the most elaborate to the simplest, we will present the methods relying on the density functional theory (DFT), the *ab initio* Hartree–Fock method and the self-consistent tight-binding approach.

Density functional methods   The density functional theory reduces the search of the many-body ground state to the resolution of an effective one-electron variational Hamiltonian, which includes a Hartree potential, which is a function of the mean charge distribution, an exchange term, due to the Pauli's principle and a correlation term.

The local density approximation (LDA) replaces the exchange and correlation potential, which is non-local, by a local potential. This latter is constructed from an approximate expression of the exchange-correlation energy $E_{xc}(\bar{n})$ of an homogeneous electron gas. At each point $\vec{r}$ of the system, the uniform electron density $\bar{n}$ in $E_{xc}(\bar{n})$ is functionally replaced by the actual density $n(\vec{r})$. The Hamiltonian and the Poisson's equation are then solved in a self-consistent manner. There have been various applications of these methods to oxides:

- the LDA–pseudo-potential method, which allowed a calculation of the electronic structure of MgO, α-quartz and rutile TiO_2 (Chang and Cohen, 1984; Glassford *et al.*, 1990; Chelikowsky *et al.*, 1991),
- the LDA–OLCAO method (OLCAO = orthogonalized linear combination of atomic orbitals) used for Cu_2O, Y_2O_3, $YBa_2Cu_3O_7$, etc (Ching, 1990),
- the LDA–PFS method (PFS = pseudo-function energy band method) applied to MgO and Al_2O_3 (French *et al.*, 1990),
- the APW method (APW = augmented plane waves) used, for example, for $SrTiO_3$ (Mattheiss, 1972),
- the DV-Xα method (DV-Xα = discrete variational-Xα method) often used in the past to simulate oxide crystals by clusters involving a few atoms (Tsukada *et al.*, 1983), at a time when the importance of screening effects and long-range delocalization was not yet recognized.

The calculations based on the DFT theory yield the band dispersion, the charge density around the atoms, and the gap width if states are expanded on an over-complete basis set. Yet, in the standard LDA approximation, the calculated band gaps are always narrower than measured ones. New methods for treating the excited states better, such as the so-called GW approximation, are now being developed to overcome this discrepancy (Hybertsen and Louie, 1986; Godby *et al.*, 1988; von der Linden and Horsch, 1988; Hott, 1991; Bechstedt *et al.*, 1992; Del Sole *et al.*, 1994), but, aside from NiO (Aryasetiawan *et al.*, 1994) and $LiNbO_3$ (Gu and

Ching, 1994; Ching *et al.*, 1994), their use has, until now, been restricted to semi-conductors. In the density functional methods the total energy is also calculated and, provided that a search of the energy minimum as a function of structural degrees of freedom is simultaneously performed, the gound state atomic configuration at 0 K may be determined.

The ab initio Hartree–Fock method At variance with the DFT methods, the *ab initio* Hartree–Fock method neglects the correlation effects. The effective one-electron Hamiltonian includes only the Hartree and exchange terms. For closed shell systems, such as most simple oxides, the ground state wave function is chosen as a Slater determinant built up with one-electron orbitals. These orbitals, which are determined in the course of the self-consistent procedure, are generally expanded on an over-complete basis set of optimized variational functions, including completely and partially filled orbitals and, often, some additional polarization orbitals. All electrons are treated at the quantum level (Silvi *et al.*, 1991; Pisani *et al.*, 1992). The *ab initio* Hartree–Fock method yields the band structure, charge density maps, the gap width and the total energy. The calculated gap widths are always much larger than the experimental values. For example, in α-alumina, Causà *et al.* (1989) find 19 eV instead of 8 eV. The *ab initio* Hartree–Fock method is more flexible and less time consuming than the LDA method. It is of valuable help e.g. in clean surface studies and adsorption phenomena. It may be coupled with a geometry optimization code, to determine the equilibrium atomic configurations.

The self-consistent tight-binding method At the lowest level of sophistication, the self-consistent tight-binding methods treat the interactions between electrons at a Hartree–Fock level, but only the outer levels, which take part in the chemical bonding, are involved in the Hamiltonian diagonalization (Harrison, 1980; Majewski and Vogl, 1986; Evarestov and Veryazov, 1991). The inner electrons contribute to the core electrostatic potential. The one-electron wave functions are expanded on an atomic orbital basis. Various techniques may be distinguished according to whether the basis set wave functions are taken under an explicit orthogonalized (CNDO = complete neglect of differential overlap) or non-orthogonalized (INDO = intermediate neglect of differential overlap, MINDO = modified INDO) form (Pople and Segal, 1965a; 1965b; Pople *et al.*, 1965a; 1965b), or under a non-explicit form (Harrison, 1980). In the first case, the empirical parameters fix the spatial dependence of the basis set wave functions. They are then used to calculate the diagonal and non-diagonal Hamiltonian matrix elements: overlap integrals, resonance integrals and coulomb terms. In the second case, the Hamiltonian matrix elements are directly parametrized and the charge dependence of the non-diagonal elements is generally neglected. The quality of the results obtained by these two approaches is similar, provided that the parametrization is well done. The empirical parameters are adjusted so as to fit the bulk electronic structure, with the constraint that the atomic orbital energies $\epsilon^0_{A,C}$ and the intra-atomic electron–electron repulsion integrals $U_{A,C}$ remain close to their tabulated values (Harrison, 1980; 1985). These methods thus cannot predict bulk properties. They have been used to study the relative stability of crystal structures (Majewski and Vogl, 1986) and to compare the electronic structure of chemically close compounds (Evarestov and Veryazov, 1990; 1991; Evarestov *et al.*, 1994). The simplicity of these methods and the ease of interpreting their results make them useful in discussing the nature of the chemical bond in complex oxides or geometries.

To emphasize the importance of the self-consistent procedure in the treatment of the electron–electron interactions in oxides, let us consider the diagonal terms of the Hamiltonian, on an atomic orbital basis set, in a Hartree approximation. For an anion, $\langle A\lambda|H|A\lambda \rangle$ reads:

$$\epsilon_{A\lambda} = \epsilon^0_{A\lambda} + U_A|Q_A| - V_A \,, \qquad (1.4.1)$$

and, for a cation, $\langle C\lambda|H|C\lambda \rangle$ is equal to:

$$\epsilon_{C\lambda} = \epsilon^0_{C\lambda} - U_C Q_C - V_C \,. \qquad (1.4.2)$$

In these expressions, the $\epsilon^0_{A\lambda,C\lambda}$ are the energies of the atomic orbitals $|A\lambda\rangle$ and $|C\lambda\rangle$ in the neutral atoms, the $Q_{A,C}$ are the ionic charges and the $U_{A,C}$ are the intra-atomic electron–electron repulsion integrals. The electrostatic potentials $V_{A,C}$ exerted by the lattice charges are also function of the ionic charges ($V_A > 0$, $V_C < 0$). In the Hartree–Fock approximation, the diagonal terms have a slightly more complex expression but they behave, as a function of the charge, in a qualitatively similar way. The atomic energy correction is associated with a modification of the wave functions: as the cation charge increases, its electronic levels are shifted towards lower energies and the atomic wave functions are more localized around the nucleus. A contraction of the ionic radii results, as already noted in Section 1.1. In the following, we will call the diagonal matrix elements $\epsilon_{A\lambda}$ and $\epsilon_{C\lambda}$ *renormalized* or effective atomic orbital energies. With respect to $\epsilon^0_{A\lambda,C\lambda}$, they have two corrections. The first, the intra-atomic correction, is related to the excess – for an anion – or to the loss – for a cation – of electron–electron repulsion on the atom under consideration. The second, due to the electrostatic potential exerted by the neighbouring ions, is called the Madelung potential. It lowers the anion levels and raises the cation levels, while the intra-atomic term acts in the opposite direction. Both corrections thus compete with each other and the balance is controlled by the value of the optical dielectric constant (Chapter 4).

The Madelung potentials can be unequivocally derived from a knowledge of the Madelung constant α only for simple structures in which anions and cations occupy equivalent lattice positions. For example, in a rocksalt crystal, with the anion and cation charges equal, respectively, to: $Q_A = -Q$ and $Q_C = Q$ ($Q > 0$), the Madelung potentials V_A and V_C read:

$$V_A = \frac{\alpha Q}{R} , \qquad (1.4.3)$$

and:

$$V_C = -\frac{\alpha Q}{R} . \qquad (1.4.4)$$

In more complex structures such as quartz or rutile, such simple relationships no longer exist. Yet, in an oxide of MO_m stoichiometry, V_A, V_C and α obey the equation:

$$V_A - V_C = -\frac{2\alpha Q_A}{mR} . \qquad (1.4.5)$$

The derivation of (1.4.5) makes use of the Madelung constant definition, (1.1.2), and of the fact that the Madelung energy E_M is equal to $(mQ_A V_A + Q_C V_C)/2$. The following are some quoted values of the Madelung potentials, calculated under the assumption that the oxides are fully ionic (Broughton and Bagus, 1980):

Na_2O :	$V_C = -10.6$ V	$V_A = 19.6$ V
ZnO :	$V_C = -24$ V	$V_A = 24$ V
MgO :	$V_C = -23.9$ V	$V_A = 23.9$ V
Al_2O_3 :	$V_C = -36.6$ V	$V_A = 26.4$ V
TiO_2 :	$V_C = -44.7$ V	$V_A = 25.8$ V
SiO_2 :	$V_C = -48.4$ V	$V_A = 30.8$ V.

The relative strengths of V_A and V_C depend upon the oxide stoichiometry. In absolute value, $|V_C|$ is lower (or larger) than V_A when Q_C is lower (or larger) than $|Q_A|$. For a given cation charge, $|V_C|$ and V_A increase as the cation ionic radius decreases, i.e. as the first neighbour inter-atomic distance R decreases.

Applications of the quantum methods Although the quantum methods require a much larger computational effort than classical approaches, they have now been applied to as wide a range of systems as the latter. For example, they have been used to describe not only the cohesive properties, such as the bulk structural parameters, the bulk modulus and the electronic structure (Binggeli *et al.*, 1991; Dovesi *et al.*, 1992; Mackrodt *et al.*, 1993; Gillan *et al.*, 1994; Takada *et al.*, 1995), but also structural phase transitions under pressure (Catti *et al.*, 1994; Liu *et al.*, 1994), defects: substitutional impurities, vacancies, etc. (De Vita *et al.*, 1992a; 1992b; 1992c; Freyria-Fava *et al.*, 1993; Sulimov *et al.*, 1994; Stefanovich *et al.*, 1994; Orlando *et al.*, 1994), and magnetism (Towler *et al.*, 1994; Catti *et al.*, 1995).

The alternating lattice model

To stress more precisely the factors which control the electronic character-istics of insulating oxides, we develop in this section a theoretical model suited to binary oxides of stoichiometry M_nO_m, in which the atoms occupy the sites of an alternating lattice (Bensoussan and Lannoo, 1979; Julien *et al.*, 1989; Goniakowski and Noguera, 1994b). It relies on a tight-binding approach, in which the eigenfunctions of the Hamiltonian are developed as linear combinations of atomic orbitals on an orthonormal basis set. Two basic assumptions underlie the model:

- All the cation (anion) outer levels have the same energy ϵ_C (ϵ_A); their degeneracy is denoted d_C (d_A). The crystal field splitting is thus neglected, and so is the energy difference between different types of atomic orbitals which may be involved in the chemical bond (e.g. 3d and 4s for titanium in TiO_2).
- The electrons can only hop from one atom onto an atom of opposite type located in its first coordination shell. Hopping processes thus connect one sub-lattice to the other as a result of the alternating character of the lattice. The oxygen-oxygen hopping processes, for example, cannot be accounted for by the model.

Again, (1.3.14), the Hamiltonian H can be split into two parts: a diagonal term H_D, which involves the site energies ϵ_A and ϵ_C, and a non-diagonal term H_{ND} associated with the resonance integrals. The eigenfunctions $|\psi_{\vec{k}}\rangle$:

$$H|\psi_{\vec{k}}\rangle = E_{\vec{k}}|\psi_{\vec{k}}\rangle , \qquad (1.4.6)$$

are equal to the sum of two components $|\psi_{\vec{k}A}\rangle$ and $|\psi_{\vec{k}C}\rangle$, which represent the projection of $|\psi_{\vec{k}}\rangle$ on the anion and cation sub-lattices. When projected on $|\psi_{\vec{k}A}\rangle$ and $|\psi_{\vec{k}C}\rangle$, (1.4.6) yields two coupled equations:

$$H_{ND}|\psi_{\vec{k}A}\rangle = (E_{\vec{k}} - \epsilon_C)|\psi_{\vec{k}C}\rangle , \qquad (1.4.7)$$

and:

$$H_{ND}|\psi_{\vec{k}C}\rangle = (E_{\vec{k}} - \epsilon_A)|\psi_{\vec{k}A}\rangle , \qquad (1.4.8)$$

which, after a second application of H_{ND}, lead to the effective Schrödinger equations:

$$H_{ND}^2|\psi_{\vec{k}A}\rangle = (E_{\vec{k}} - \epsilon_A)(E_{\vec{k}} - \epsilon_C)|\psi_{\vec{k}A}\rangle \,, \tag{1.4.9}$$

and:

$$H_{ND}^2|\psi_{\vec{k}C}\rangle = (E_{\vec{k}} - \epsilon_A)(E_{\vec{k}} - \epsilon_C)|\psi_{\vec{k}C}\rangle \,. \tag{1.4.10}$$

The resolution of (1.4.6) is thus equivalent to finding the eigenfunctions $|\psi_{\vec{k}A}\rangle$ and $|\psi_{\vec{k}C}\rangle$ and eigenvalues $F_{\vec{k}} = (E_{\vec{k}} - \epsilon_A)(E_{\vec{k}} - \epsilon_C)$ of H_{ND}^2 on one sub-lattice. Due to the positive definite character of the operator H_{ND}^2, all the $F_{\vec{k}}$ values are larger than or equal to zero.

For a given Bloch vector $\vec{k}$, the degeneracy of $F_{\vec{k}}$ is nd_C on the cation sub-lattice and md_A on the anion one. When nd_C is different from md_A, there are two types of solutions for $\psi_{\vec{k}}$: there are n_0 non-trivial solutions ($n_0 = $ min. (nd_C, md_A)) and n_1 solutions ($n_1 = $ max. $(nd_C, md_A) - n_0$) for which either $\psi_{\vec{k}A}$ (if $md_A < nd_C$) or $\psi_{\vec{k}C}$ (if $md_A > nd_C$) identically vanishes. The eigenenergies for the latter are equal to ϵ_C or ϵ_A, respectively. In order to be specific, we will assume in the following that $md_A < nd_C$ and that the diagonalization of H_{ND}^2 is performed on the anion sub-lattice.

Once the solutions of H_{ND}^2 are obtained, i.e. after the determination of the band dispersion $F_{\vec{k}}$, of the range of existence of $F_{\vec{k}}$: $[F_{min}, F_{max}]$ and of the density of states $M(F)$, it is possible to deduce the total density of states $N(E)$ and the local densities of states $N_A(E)$ and $N_C(E)$ on the anion and cation sub-lattices, thanks to the following relationships (δ is the Kronecker symbol):

$$N(E) = 2\left|E - \frac{(\epsilon_C + \epsilon_A)}{2}\right| M(E) + n_1\delta(E - \epsilon_C) \,, \tag{1.4.11}$$

and:

$$N_A(E) = |E - \epsilon_C| M(E) \,, \tag{1.4.12}$$

$$N_C(E) = |E - \epsilon_A| M(E) + n_1\delta(E - \epsilon_C) \,. \tag{1.4.13}$$

One has used the equality $F = (E - \epsilon_A)(E - \epsilon_C)$ in the expression for $N(E)$, and $M(E)$ is the transcription of $M(F)$ by the change of variable F into E. In the calculation of $N_A(E)$ and $N_C(E)$, the weight of $\psi_{\vec{k}}$ on each sub-lattice is estimated thanks to:

$$\langle\psi_{\vec{k}C}|H_{ND}|\psi_{\vec{k}A}\rangle = (E_{\vec{k}} - \epsilon_C)\langle\psi_{\vec{k}C}|\psi_{\vec{k}C}\rangle = (E_{\vec{k}} - \epsilon_A)\langle\psi_{\vec{k}A}|\psi_{\vec{k}A}\rangle \,, \tag{1.4.14}$$

and by using the normalization condition:

$$\langle\psi_{\vec{k}}|\psi_{\vec{k}}\rangle = \langle\psi_{\vec{k}A}|\psi_{\vec{k}A}\rangle + \langle\psi_{\vec{k}C}|\psi_{\vec{k}C}\rangle = 1 \,. \tag{1.4.15}$$

To each eigenvalue $F_{\vec{k}}$ are associated two energies $E_{\vec{k}}^{\pm}$ equal to:

$$E_{\vec{k}}^{\pm} = \frac{(\epsilon_C + \epsilon_A)}{2} \pm \frac{1}{2}\sqrt{(\epsilon_A - \epsilon_C)^2 + 4F_{\vec{k}}} \ . \tag{1.4.16}$$

The '$-$' sign refers to the valence band and the '$+$' sign to the conduction band. Due to the factor $|E - \epsilon_C|$ in (1.4.12), the weight of $N_A(E)$ in the conduction band is lower than in the valence band. The reverse is true for $N_C(E)$. This property is shared by all insulating oxides: the valence band is mainly built of oxygen orbitals, while the cation states mainly determine the conduction band. It is exemplified in Fig. 1.6 which displays the local densities of states on the magnesium and oxygen orbitals in MgO (Russo and Noguera, 1992a). According to (1.4.12), $N_A(E)$ vanishes for $E = \epsilon_C$. This energy is located at the bottom of the conduction band, if $F_{\min} = 0$, and the state has a cation non-bonding character. At the top of the valence band, similarly, $N_C(E)$ vanishes and the eigenstate has an anion non-bonding character.

The alternating lattice method is easy to use on all simple or more complex lattices, when the atomic orbitals have an s character, because the resonance integrals β are isotropic. For example, in an alternating linear chain, with cells of size a, indexed by the integer n, the application of H_{ND} to an anion orbital gives: $H_{ND}|A_n\rangle = \beta(|C_n\rangle + |C_{n+1}\rangle)$, i.e.:

$$H_{ND}^2|A_n\rangle = 2\beta^2|A_n\rangle + \beta^2\left(|A_{n+1}\rangle + |A_{n-1}\rangle\right) \ . \tag{1.4.17}$$

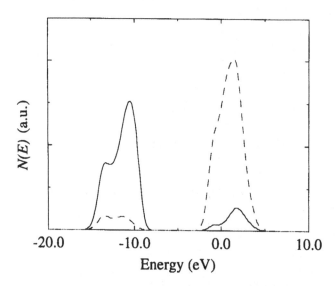

Fig. 1.6.   MgO density of states, projected on the oxygen and magnesium (full and dashed lines, respectively) orbitals (in arbitrary units).

Expression (1.4.17) is reminiscent of the result on a chain of identical atoms, with atomic energies equal to $2\beta^2$ and effective nearest neighbour resonance integrals β^2. The eigenvalues are thus:

$$F_k = 2\beta^2 + 2\beta^2 \cos ka . \qquad (1.4.18)$$

The extrema of F_k are reached at the zone edge and centre: $F_{min} = 0$ and $F_{max} = 4\beta^2$. Within the assumption of s orbitals, we give in the following the expressions of $F_{\vec{k}}$, F_{min} and F_{max} for several simple lattices. For each of them, we note the cation coordination number Z:

- Alternating linear chain (lattice parameter a; $Z = 2$):

$$F_k = 2\beta^2 + 2\beta^2 \cos ka$$
$$F_{min} = 0 \; ; \; F_{max} = 4\beta^2 . \qquad (1.4.19)$$

- Alternating square lattice (lattice parameter a; $Z = 4$):

$$F_{\vec{k}} = 4\beta^2 \left(1 + \cos k_x a + \cos k_y a + \cos k_x a \cos k_y a\right)$$
$$F_{min} = 0 \; ; \; F_{max} = 16\beta^2 . \qquad (1.4.20)$$

- Three-dimensional CsCl lattice (lattice parameter a; $Z = 8$):

$$F_{\vec{k}} = 8\beta^2 \left(1 + \cos k_x a + \cos k_y a + \cos k_z a\right)$$
$$+ 8\beta^2 \left(\cos k_x a \cos k_y a + \cos k_y a \cos k_z a + \cos k_z a \cos k_x a\right)$$
$$F_{min} = 0 \; ; \; F_{max} = 64\beta^2 . \qquad (1.4.21)$$

- Three-dimensional NaCl lattice (lattice parameter of the large cubic cell a; $Z = 6$):

$$F_{\vec{k}} = \beta^2 \left[6 + 2 \left(\cos k_x a + \cos k_y a + \cos k_z a\right)\right]$$
$$+ 8\beta^2 \left(\cos k_x \frac{a}{2} \cos k_y \frac{a}{2} + \cos k_y \frac{a}{2} \cos k_z \frac{a}{2} + \cos k_z \frac{a}{2} \cos k_x \frac{a}{2}\right)$$
$$F_{min} = 0 \; ; \; F_{max} = 36\beta^2 . \qquad (1.4.22)$$

- Three-dimensional ZnS blende lattice (lattice parameters a; $b = a\sqrt{2}/2$; $Z = 4$):

$$F_{\vec{k}} = 4\beta^2 \left(1 + \cos k_x b \cos k_y b + \cos k_y b \cos k_z b + \cos k_z b \cos k_x b\right)$$
$$F_{min} = 0 \; ; \; F_{max} = 16\beta^2 . \qquad (1.4.23)$$

- Three-dimensional rutile lattice TiO_2 (lattice parameters $a = b \neq c$; $Z = 6$; the resonance integrals for the two Ti–O inter-atomic distances are assumed to be equal):

$$F_{\vec{k}} = \beta^2 \left(6 + 8 \cos k_x \frac{a}{2} \cos k_y \frac{b}{2} \cos k_z \frac{c}{2} + 4 \cos k_z c\right)$$
$$F_{min} = 0 \; ; \; F_{max} = 18\beta^2 . \qquad (1.4.24)$$

• In a planar honeycomb lattice or in a three-dimensional diamond lattice, hybrid sp$_2$ or sp$_3$ atomic orbitals are eigenfunctions of H^2_{ND}, with eigenvalues β^2. If one assumes that they are orthogonal and degenerate, and if hopping processes take place only between orbitals pointing along the bond direction, the eigenvalues are independent of the Bloch wave vector $\vec{k}$:

$$F_{\vec{k}} = \beta^2$$
$$F_{min} = F_{max} = \beta^2 \ . \qquad\qquad (1.4.25)$$

Although oversimplified, this situation exemplifies a case in which F_{min} is non-zero.

When the orbital symmetry allows several orbital overlaps (σ and π, for example, for p orbitals), it is not generally possible to derive analytically the complete band structure $F_{\vec{k}}$. Yet, the alternating lattice approach still gives information on the band limits and on the gap width. In addition, it also provides expressions for the ionic charges and the covalent energy if a simple analytical form of $M(F)$ is assumed. We will develop these different points in the following paragraphs.

Band limits and band widths

We distinguish the band spreading, which is the energy range in which the density of states is non-zero, from the band width, which is an average quantity related to the second moment of the density of states.

Band limits   The energies at the bottom of the valence band and at the top of the conduction band, i.e. the smallest $E^-_{\vec{k}}$ value and the largest $E^+_{\vec{k}}$ value, are associated with F_{max} in (1.4.16). On the other hand, the gap edges, i.e. the largest $E^-_{\vec{k}}$ value and the smallest $E^+_{\vec{k}}$ value, are a function of F_{min}.

On a given sub-lattice, the eigenfunction of H^2_{ND} which corresponds to F_{max} has the smallest number of nodes. Due to the sign of the effective resonance integrals β^2, it has the strongest anti-bonding character. On the full lattice, it corresponds either to a bonding anion–cation state at the bottom of the valence band, or to an anti-bonding state at the top of the conduction band. In a similar way, F_{min} is associated with the strongest bonding state on one of the sub-lattices, i.e. to non-bonding cation or anion states on the full lattice, when symmetry properties allow it. These remarks are illustrated on Fig. 1.7, for an alternating square lattice involving s orbitals.

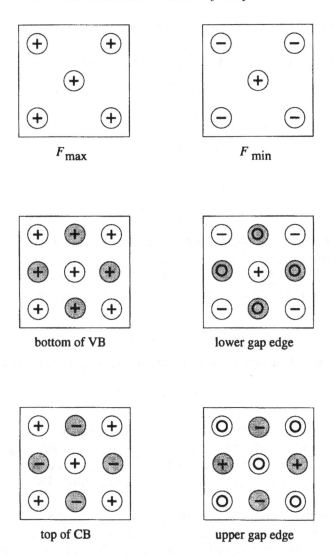

Fig. 1.7. Phases of the wave functions at the band limits, for an alternating two-dimensional square lattice involving s orbitals. The top part of the figure represents the phases on one of the sub-lattices, and the two lower parts show the corresponding phases on the full lattice. On the left hand side, are the states associated with F_{max}: depending upon the sign of the wave function on the second sub-lattice, its number of nodes is either vanishing or maximal, which gives the bottom of the valence band (VB) and the top of the conduction band (CB), respectively. On the right hand side, are the states associated with F_{min}: the coefficients of the wave function on one sub-lattice are equal to zero and are out of phase on the other sub-lattice; these non-bonding states are located at the lower and upper gap edges.

According to (1.4.16), the valence band ranges from E_{min}^- to E_{max}^-, while the conduction band is located in the energy range $[E_{min}^+, E_{max}^+]$. The valence and conduction bands have thus the same spreading, equal to:

$$\Delta E = \frac{1}{2}\sqrt{(\epsilon_C - \epsilon_A)^2 + 4F_{max}} - \frac{1}{2}\sqrt{(\epsilon_C - \epsilon_A)^2 + 4F_{min}} \,. \qquad (1.4.26)$$

ΔE approaches:

$$\Delta E = \frac{F_{max} - F_{min}}{(\epsilon_C - \epsilon_A)} \,, \qquad (1.4.27)$$

if ionic effects prevail, and

$$\Delta E = \sqrt{F_{max}} - \sqrt{F_{min}} \,, \qquad (1.4.28)$$

otherwise. To be more specific, let us consider an oxide MO_m, described with s orbitals and a cation coordination number equal to Z. The generalization of (1.4.19) to (1.4.24) gives: $F_{min} = 0$ and $F_{max} = Z^2\beta^2/m$. In the ionic limit, the band spreading is proportional to the square of $Z\beta$:

$$\Delta E = \frac{Z^2\beta^2}{m(\epsilon_C - \epsilon_A)} \,, \qquad (1.4.29)$$

while in the covalent limit, it is proportional to $Z\beta$, as in metals:

$$\Delta E = \frac{Z\beta}{\sqrt{m}} \,. \qquad (1.4.30)$$

These relationships exemplify how the band edges are sensitive to the anion–cation electron delocalization. Second-neighbour hopping broadens the bands, but in a way which cannot be predicted by the alternating lattice model, because it hinders the Hamiltonian decoupling, (1.4.7) and (1.4.8).

First moments of the local densities of states A model cannot account for all the details of the actual density of states but it can predict reasonable values for some integral quantities, such as the first moments. One defines the ith moment M_i of $N(E)$ by:

$$M_i = \int E^i N(E) dE \,. \qquad (1.4.31)$$

The three first moments have a simple physical meaning. M_0 is the band degeneracy; M_1/M_0 gives the mean energy of the band, i.e. its centre of gravity; and the centred second moment:

$$M_2' = \int \left(E - \frac{M_1}{M_0}\right)^2 N(E) dE \,, \qquad (1.4.32)$$

once normalized to M_0, is equal to the square of the band width. One should note that the energy summation in (1.4.31) and (1.4.32) includes the valence *and* conduction states. For insulators or semi-conductors, M_1/M_0 and M_2'/M_0 thus describe the average position and the width of the (valence + conduction) bands. In the simplest metallic case, in which the atoms are of a single type, involve s orbitals of energy ϵ_0 and are surrounded by Z first neighbours, the first moments have the following values: $M_1/M_0 = \epsilon_0$ and $M_2'/M_0 = Z\beta^2$. When inequivalent sites are present in a system, because of the presence of a surface or of several types of atoms, it is interesting to also consider the moments of the *local* densities of states projected on a given site and, if necessary, on a given orbital momentum.

We now derive the expressions for the two first moments M_1/M_0 and M_2'/M_0 of the local density of states (LDOS), normalized to one atom and one spin direction, for an oxide M_nO_m, in a tight-binding approach. For this purpose, we will not need the two basic assumptions of the alternating lattice model. We will note $|A\lambda\rangle$ and $|C\mu\rangle$ the anion and cation atomic orbitals, and $\epsilon_{A\lambda}$ and $\epsilon_{C\mu}$ their energies. The basis set is assumed to be orthonormal. The first moment M_1 of an anion LDOS reads:

$$
\begin{aligned}
M_{1A} &= \int E N_A(E) dE \\
&= \sum_{\vec{k}\lambda} \int E \left| \langle A\lambda | \psi_{\vec{k}} \rangle \right|^2 \delta(E - E_{\vec{k}}) dE \\
&= \sum_{\vec{k}\lambda} E_{\vec{k}} \left| \langle A\lambda | \psi_{\vec{k}} \rangle \right|^2 .
\end{aligned}
\tag{1.4.33}
$$

Since the $\psi_{\vec{k}}$ are the eigenfunctions of H with eigenenergies $E_{\vec{k}}$, M_{1A} may also be written:

$$
M_{1A} = \sum_{\vec{k}\lambda} \langle A\lambda | H | \psi_{\vec{k}} \rangle \langle \psi_{\vec{k}} | A\lambda \rangle .
\tag{1.4.34}
$$

It may be simplified by using the closure relation $\sum_{\vec{k}} |\psi_{\vec{k}}\rangle\langle\psi_{\vec{k}}| = I$:

$$
M_{1A} = \sum_{\lambda} \langle A\lambda | H | A\lambda \rangle ,
\tag{1.4.35}
$$

which gives the final result:

$$
\frac{M_{1A}}{M_{0A}} = \frac{1}{d_A} \sum_{\lambda} \epsilon_{A\lambda} .
\tag{1.4.36}
$$

Similarly:

$$
\frac{M_{1C}}{M_{0C}} = \frac{1}{d_C} \sum_{\mu} \epsilon_{C\mu} .
\tag{1.4.37}
$$

The centres of gravity of $N_A(E)$ and $N_C(E)$ are thus located at the mean atomic orbital energy.

The derivation of the centred second moment of $N_A(E)$ is very similar, if one realizes that the closure relation may either be written on the eigenfunction $|\psi_{\vec{k}}\rangle$ or on the atomic orbital $|J\lambda'\rangle$ basis sets:

$$M'_{2A} = \int \left(E - \frac{1}{d_A}\sum_\lambda \epsilon_{A\lambda}\right)^2 N_A(E)dE$$

$$= \sum_{\vec{k}\lambda} \int \left(E - \frac{1}{d_A}\sum_\lambda \epsilon_{A\lambda}\right)^2 dE \left|\langle A\lambda|\psi_{\vec{k}}\rangle\right|^2 \delta(E - E_{\vec{k}})$$

$$= \sum_{\vec{k}\lambda} \left(E_{\vec{k}} - \frac{1}{d_A}\sum_\lambda \epsilon_{A\lambda}\right)^2 \left|\langle A\lambda|\psi_{\vec{k}}\rangle\right|^2$$

$$= \sum_{\vec{k}\lambda} \left\langle A\lambda\left|H - \frac{1}{d_A}\sum_\lambda \epsilon_{A\lambda}\right|\psi_{\vec{k}}\right\rangle\left\langle \psi_{\vec{k}}\left|H - \frac{1}{d_A}\sum_\lambda \epsilon_{A\lambda}\right|A\lambda\right\rangle$$

$$= \sum_{\lambda J\lambda'} \left\langle A\lambda\left|H - \frac{1}{d_A}\sum_\lambda \epsilon_{A\lambda}\right|J\lambda'\right\rangle\left\langle J\lambda'\left|H - \frac{1}{d_A}\sum_\lambda \epsilon_{A\lambda}\right|A\lambda\right\rangle .$$

$$(1.4.38)$$

In the last step of the calculation, the diagonal and non-diagonal parts of the Hamiltonian are distinguished:

$$\frac{M'_{2A}}{M_{0A}} = \frac{1}{d_A}\sum_\lambda \left(\epsilon_{A\lambda} - \frac{1}{d_A}\sum_{\lambda'} \epsilon_{A\lambda'}\right)^2 + \frac{1}{d_A}\sum_{\lambda J\lambda'} |\langle A\lambda|H_{ND}|J\lambda'\rangle|^2 . \quad (1.4.39)$$

The band width has thus two contributions: one related to the atomic orbital energy splitting and the other to the covalent effects, i.e. to the electron delocalization between the Z_A first-neighbour anion–cation pairs, the Z'_A second-neighbour anion–anion pairs, etc. On simple lattices on which the Z_A first-neighbour inter-atomic distances are equal, the sum $\sum_{\lambda\mu} |\langle A\lambda|H_{ND}|C\mu\rangle|^2$ does not depend on C and the first-neighbour contribution is proportional to Z_A. The second-neighbour term has a similar property. The centred second moment then simply depends on: $\beta_1^2 = \sum_{\lambda\mu} |\langle A\lambda|H_{ND}|C\mu\rangle|^2/d_A$ and $\beta_2^2 = \sum_{\lambda\mu} |\langle A\lambda|H_{ND}|A\mu\rangle|^2/d_A$ which are the square of the effective first and second neighbour resonance integrals:

$$\frac{M'_{2A}}{M_{0A}} = \frac{1}{d_A}\sum_\lambda \left(\epsilon_{A\lambda} - \frac{1}{d_A}\sum_{\lambda'} \epsilon_{A\lambda'}\right)^2 + Z_A\beta_1^2 + Z'_A\beta_2^2 . \quad (1.4.40)$$

If the anion orbitals are degenerate, the first term vanishes and M'_{2A}/M_{0A} becomes formally identical to the expression of the second moment in metallic systems, under similar assumptions. The width of the LDOS is an increasing function of the coordination number, a result which will be used in Chapter 3 to discuss the modifications of the electronic structure at surfaces.

The gap width

In this paragraph, devoted to the analysis of the gap width Δ, we will use a one-electron picture in which Δ is the energy difference between the highest occupied orbital (HOMO = highest occupied molecular orbital: here the top of the valence band) and the lowest unoccupied orbital (LUMO = lowest unoccupied molecular orbital: here the bottom of the conduction band). This amounts to neglecting the excitonic effects which take place in a gap measurement. A discussion of excitons is postponed to Chapter 4.

Result of the alternating lattice model   In the classical models of insulators, the gap width Δ is equal to the energy difference between the cation and anion orbital energies, corrected by the Madelung potential:

$$\Delta = \epsilon_C - \epsilon_A . \qquad (1.4.41)$$

The alternating lattice method, developed above, allows one to include electron delocalization effects.

Under the two basic assumptions of the model, we have found that the band edges are derived from the smallest eigenvalue F_{min} of H_{ND}^2. The gap width is thus equal to:

$$\Delta = \sqrt{(\epsilon_C - \epsilon_A)^2 + 4F_{min}} , \qquad (1.4.42)$$

in the absence of uncoupled states ($nd_C = md_A$), or to:

$$\Delta = \frac{(\epsilon_C - \epsilon_A)}{2} + \sqrt{(\epsilon_C - \epsilon_A)^2 + 4F_{min}} , \qquad (1.4.43)$$

otherwise.

Equation (1.4.42) for the gap width is very similar to the expression proposed by Phillips (1970) for E_g as recalled in Section 1.3. It has two contributions: an ionicity gap ($\epsilon_C - \epsilon_A$) and a covalent gap, which, here, reads $2\sqrt{F_{min}}$. Since F_{min} depends both upon the resonance integrals β and the Bloch wave vectors $\vec{k}$ at the gap edges, F_{min} is a function of the nature and symmetry of the orbitals involved in the chemical bond, of their relative directions on neighbouring atoms and more generally of the crystal structure.

On a diamond or honeycomb lattice, described by (1.4.25), or for quartz, α-alumina, etc., in which the crystal structure and the orbital character induce non-negligible angular effects, F_{min} is non-zero.

In some other cases, despite a large value of the ratio: $\beta/(\epsilon_C - \epsilon_A)$, the gap opens at a vector of the Brillouin zone where F_{min} vanishes. This occurs on all the lattices that we have considered above, (1.4.19) to (1.4.24), when the orbitals have an s character. It also occurs for compounds with s or p orbitals in a rocksalt structure, for example, for NaCl and MgO (Goniakowski and Noguera, 1994b). The ratio $2\sqrt{F_{\min}}/(\epsilon_C - \epsilon_A)$ thus does not characterize the mixed iono–covalent character of an insulator in an accurate way. It may only be used to compare the ionicity of a series of compounds which have the same structure, provided that $F_{\min}$ is non-zero.

Anion–anion and cation–cation delocalization   As stated above, the alternating lattice model cannot account for electron events taking place inside a given sub-lattice, because they invalidate (1.4.7) and (1.4.8). Numerical calculations prove that such processes have two consequences. On the one hand, they widen the bands: in oxides, oxygen–oxygen hopping is generally required to account for the valence band widths, because oxygen atoms are nearly in contact. On the other hand, they induce a gap narrowing: the band spreading increases at the expense of the gap.

To summarize, within the assumptions of the alternating lattice model, the gap width has two contributions related to the anion–cation difference in electronegativity and to covalent effects. The covalent contribution is an increasing function of the resonance integrals, but it also depends upon the vectors at which the gap opens and closes in reciprocal space, i.e. upon the symmetry of the orbitals and of the lattice. In specific cases, the orbital crystal field splitting and second-neighbour delocalization effects may slightly modify this simple picture.

Ionic charges

We now turn to a discussion of the parameters which fix the values of the ionic charges. As already discussed above, the purpose is not to obtain quantitative estimates for the bulk charges in a given oxide, because these quantities remain controversial in the literature. Different authors obtain different estimates of the ionicity for the same compound, e.g. in Al_2O_3 (Causà *et al.*, 1989; Guo *et al.*, 1992), or even different trends in a given series, e.g. the alkaline-earth oxide series from MgO to BaO (Pacchioni *et al.*, 1993; Sousa *et al.*, 1993a; 1993b; Noguera *et al.*, 1993). Yet, it is interesting to point out how the anion–cation charge transfer is related to the geometric characteristics of the oxide, to the cation electronegativity and to the value of the optical dielectric constant. Although the ionic charges cannot be measured, it is worthwhile to stress the parameters upon which they depend (Catlow and Stoneham, 1983), because they characterize the ionicity of an insulator. In addition, we will see in Chapter 6 that the value of the oxygen charge is used to discuss the reactivity of oxides. This discussion will also give us the bases to understand the modifications of charge transfer at surfaces.

In the preceding paragraphs, we have considered quantities related to the valence *plus* conduction band states. The oxide ionic charges and covalent energy, on the other hand, depend only upon the occupied valence states. We now present a derivation of these two quantities, within the assumptions of the alternating lattice model. Similar theoretical methods involving a calculation of the density of states by the continued fraction method (Haydock *et al.*, 1972; 1975; Turchi *et al.*, 1982), or by the moment method (Gaspard and Cyrot-Lackmann, 1973), were used in the past for metallic systems to estimate their band energy, their surface energy, their adsorption characteristics, etc.

To obtain information on the valence band states, one has to start from an approximate expression of $M(F)$. The simplest form is a delta function peaked at the energy F of its first moment, with a weight equal to n_0 (defined p. 26). Due to the quadratic relationship between $F_{\vec{k}}$ and $E_{\vec{k}}$, the first moment of $M(F)$ on the anion sub-lattice is equal to the covalent contribution to the second moment of $N_A(E)$, i.e. $Z_A\beta^2$:

$$M(F) = n_0\delta\left(F - Z_A\beta^2\right) . \tag{1.4.44}$$

By choosing the zero of energy at $(\epsilon_C + \epsilon_A)/2$, by defining $x = \epsilon_C - \epsilon_A$ to simplify the notations and by using (1.4.11), (1.4.12) and (1.4.13), one may write the total and local densities of states in the following way:

$$N(E) = n_0\delta\left(E + \sqrt{Z_A\beta^2 + x^2/4}\right)$$
$$+ n_0\delta\left(E - \sqrt{Z_A\beta^2 + x^2/4}\right) + n_1\delta\left(E - x/2\right) . \tag{1.4.45}$$

The total density of states displays two peaks located at symmetric positions with respect to $(\epsilon_C + \epsilon_A)/2$. The same is true as regards the local densities of states $N_A(E)$ and $N_C(E)$:

$$N_A(E) = n_0\frac{\left(x + \sqrt{4Z_A\beta^2 + x^2}\right)}{2\sqrt{4Z_A\beta^2 + x^2}}\delta\left(E + \sqrt{Z_A\beta^2 + x^2/4}\right)$$
$$+ n_0\frac{\left(\sqrt{4Z_A\beta^2 + x^2} - x\right)}{2\sqrt{4Z_A\beta^2 + x^2}}\delta\left(E - \sqrt{Z_A\beta^2 + x^2/4}\right) ,$$
$$\tag{1.4.46}$$

and:

$$
N_C(E) = n_0 \frac{\left(\sqrt{4Z_A\beta^2 + x^2} - x\right)}{2\sqrt{4Z_A\beta^2 + x^2}} \delta\left(E + \sqrt{Z_A\beta^2 + x^2/4}\right)
$$
$$
+ n_0 \frac{\left(x + \sqrt{4Z_A\beta^2 + x^2}\right)}{2\sqrt{4Z_A\beta^2 + x^2}} \delta\left(E - \sqrt{Z_A\beta^2 + x^2/4}\right)
$$
$$
+ n_1 \delta(E - x/2) . \tag{1.4.47}
$$

As far as $N_A(E)$ is concerned, the valence peak is higher than the conduction peak. The reverse is true for $N_C(E)$. The centre of gravity of $N_A(E)$ is $E = \epsilon_A$, in agreement with the expression (1.4.36) of the first moment for degenerate orbitals. With respect to $E = \epsilon_A$, the valence peak of $N_A(E)$ is shifted towards lower energies by $\sqrt{Z_A\beta^2 + x^2/4} - x/2$, an increasing function of $Z_A\beta^2$. When starting from a more realistic form for $M(F)$: either a sum of two delta functions or a rectangular shape, two additional features appear. First, the valence and conduction band widths are non-zero and increase with the width of $M(F)$. Second, the centres of gravity of $N_A(E)$ and $N_C(E)$, in the valence band, are shifted relative to each other. That of $N_A(E)$ is located at a higher energy, due to the presence of anion non-bonding states at the top of the valence band, which do not involve the cation orbitals.

The integration of $N_A(E)$ in the valence band yields the anion electron number; the absolute value Q of the ionic charge is equal to:

$$
Q = 2 - \frac{n_0}{m} \left(1 - \frac{\epsilon_C - \epsilon_A}{\sqrt{(\epsilon_C - \epsilon_A)^2 + 4Z_A\beta^2}}\right) . \tag{1.4.48}
$$

This is the generalization of the expression (1.3.13), found for the diatomic molecule, to compounds with higher coordination numbers Z_A and degenerate orbitals. Q is equal to 2 when $\beta/(\epsilon_C - \epsilon_A)$ vanishes. The ratio $(\epsilon_C - \epsilon_A)^2/[(\epsilon_C - \epsilon_A)^2 + 4Z_A\beta^2]$ thus characterizes the ionicity of a mixed iono–covalent material. This conclusion is in agreement with Phillips' ionicity scale: the energy E_g that he considers is the energy difference between the centres of gravity of the valence and conduction bands. It may be estimated in the present model, using (1.4.45): $E_g = \sqrt{(\epsilon_C - \epsilon_A)^2 + 4Z_A\beta^2}$. The alternating lattice model thus gives a prescription, based on microscopic arguments, by which to calculate Phillips' hetero-polar $E_h = \epsilon_C - \epsilon_A$ and covalent $E_c = 2\beta\sqrt{Z_A}$ 'gaps'.

According to (1.4.48), Q depends upon two parameters: the number of coupled states n_0 and the ratio $\beta\sqrt{Z_A}/(\epsilon_C - \epsilon_A)$. The latter is an implicit function of Q because ϵ_C and ϵ_A are renormalized by electron–electron

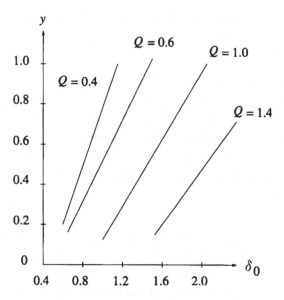

Fig. 1.8. Constant charge diagram: on the horizontal axis is shown δ_0, equal to $\left(\epsilon_C^0 - \epsilon_A^0\right)/\beta\sqrt{Z_A}$ and on the vertical axis is reported the value of y equal to $\left(U_A + U_C - 2\alpha/mR\right)/\beta\sqrt{Z_A}$.

interactions. In an Hartree approximation, (1.4.1) and (1.4.2), the energy difference $\epsilon_C - \epsilon_A$ reads:

$$\epsilon_C - \epsilon_A = \epsilon_C^0 - \epsilon_A^0 + Q\left(\frac{2\alpha}{mR} - U_A - U_C\right) , \qquad (1.4.49)$$

an expression in which we have made explicit the dependence of the Madelung potential upon the oxygen charge: $V_C - V_A = 2Q\alpha/mR$. Equations (1.4.48) and (1.4.49) are a system of coupled equations, in which two dimensionless parameters appear: $\delta_0 = \left(\epsilon_C^0 - \epsilon_A^0\right)/\beta\sqrt{Z_A}$ and $y = \left(U_A + U_C - 2\alpha/mR\right)/\beta\sqrt{Z_A}$. δ_0 characterizes the ionicity of an unpolarizable system, i.e. a system in which the atomic levels are not modified by electrostatic perturbations. The second parameter involves a competition between intra- and inter-atomic electron–electron interactions, i.e. upon the range of the coulomb potential. It is related to the value of the dielectric constant which determines the response to an alternating potential, which, in a tight-binding approach, is proportional to (see Chapter 4):

$$\epsilon - 1 \propto \frac{(2 - Q)(U_A + U_C - 2\alpha/mR)}{\epsilon_C - \epsilon_A} . \qquad (1.4.50)$$

Strong screening can be correlated to large and positive values of the factor $U_A + U_C - 2\alpha/mR$. Fig. 1.8 represents the constant charge lines obtained by self-consistently solving (1.4.48) and (1.4.49) as a function of δ_0, shown on the horizontal axis, and y, shown on the vertical axis. It proves that:

- the larger the energy difference $\left(\epsilon_C^0 - \epsilon_A^0 \right)$, the larger the charge,
- for constant values of y, the smaller the coordination number Z_A and the strength of the resonance integrals β, the larger the charge,
- the smaller the polarizability, i.e. the smaller the oxide optical dielectric function, the larger the charge.

The dimensionless ratios: δ_0 and y determine *in fine* the oxide ionicity.

Cohesion energy

In a self-consistent treatment of the electrostatic interactions, the cohesion energy contains four contributions. We will first calculate them for a diatomic molecule, i.e. we will generalize the derivation made in Section 1.3. We will account for electron–electron interactions in an Hartree approximation, (1.4.1) and (1.4.2). The electrostatic potentials V_A and V_C exerted on each atom by the neighbouring charge $Q_{C,A}$, located at a distance R, read (in atomic units):

$$V_{A,C} = \frac{Q_{C,A}}{R} . \tag{1.4.51}$$

The charges are functions of the numbers of valence electrons: N_A and N_C in the molecular state and N_A^0 and N_C^0 in the neutral atoms ($Q_C = N_C^0 - N_C$), and thus:

$$\epsilon_A = \epsilon_A^0 + U_A \left(N_A - N_A^0 \right) + \frac{\left(N_C - N_C^0 \right)}{R} , \tag{1.4.52}$$

and:

$$\epsilon_C = \epsilon_C^0 + U_C \left(N_C - N_C^0 \right) + \frac{\left(N_A - N_A^0 \right)}{R} . \tag{1.4.53}$$

The electronic energy of the AC molecule is equal to the sum of the energies of the occupied molecular states – here, twice the bonding state energy – minus the electron–electron interactions which have been counted twice:

$$E_{el} = 2E_L - \frac{U_A N_A N_A}{2} - \frac{U_C N_C N_C}{2} - \frac{N_A N_C}{R} . \tag{1.4.54}$$

Two additional terms are required to get the total energy E: a coulomb repulsion energy between the ionic cores, which is equal to $N_A^0 N_C^0 / R$, and a short-range repulsion energy, already discussed in Born's model. In numerical approaches which account for atomic orbital overlap, this latter term is already included among the exchange terms and it is not necessary to add it. We will call it E_{rep}:

$$E = 2E_L - \frac{U_A N_A N_A}{2} - \frac{U_C N_C N_C}{2} + \frac{\left(N_A^0 N_C^0 - N_A N_C\right)}{R} + E_{rep} . \quad (1.4.55)$$

As previously, $2E_L$ contains the Hamiltonian non-diagonal E_{ND} and diagonal E_D contributions. The expression of E_D:

$$E_D = N_A \epsilon_A + N_C \epsilon_C , \quad (1.4.56)$$

differs from (1.3.15) because of the renormalization of the atomic energies. One deduces:

$$N_A \epsilon_A = N_A \left(\epsilon_A^0 - U_A N_A^0\right) + U_A N_A N_A + \frac{N_A \left(N_C - N_C^0\right)}{R} . \quad (1.4.57)$$

The intra-atomic and inter-atomic electron-electron interactions which were counted twice: $U_A N_A N_A$ and $N_A N_C / R$, appear clearly in (1.4.57). Once all the terms are gathered, the total energy E reads:

$$E = E_{int} + E_{cov} + E_M + E_{rep} . \quad (1.4.58)$$

The internal energy:

$$E_{int} = N_A \left(\epsilon_A^0 - U_A N_A^0\right) + \frac{U_A N_A N_A}{2}$$
$$+ N_C \left(\epsilon_C^0 - U_C N_C^0\right) + \frac{U_C N_C N_C}{2} , \quad (1.4.59)$$

accounts for the atomic orbital filling and for the intra-atomic repulsion between electrons. The covalent energy E_{cov} is a function of the resonance integral β and the charge Q:

$$E_{cov} = -2|\beta| \sqrt{\delta Q(Q_{max} - \delta Q)} , \quad (1.4.60)$$

where δQ, represents the deviation from the ionic limit (here $Q_{max} = 1$). E_{cov} characterizes the bond covalency. This expression will be used in Chapter 6 to discuss the relationship between the structural properties and the oxygen–proton charge transfer of hydroxyl groups adsorbed on oxide surfaces. The electrostatic interaction between the two ionic charges at a distance R is the Madelung energy E_M:

$$E_M = \frac{\left(N_A - N_A^0\right)\left(N_C - N_C^0\right)}{R} . \quad (1.4.61)$$

Finally, E_{rep} is the short-range repulsion energy.

In a crystal, the expressions for E_{int}, E_{cov}, E_M and E_{rep} are more involved, but the principle of the decomposition still applies. The same hypotheses as those used in the calculation of the ionic charges yield an expression for the covalent energy, in a crystal, by a mere integration of $EN(E)$ in the valence energy range, and a substraction of the filling energy $N_A \epsilon_A + N_C \epsilon_C$:

$$E_{\text{cov}} = \frac{-4n_0 Z_A \beta^2}{\sqrt{(\epsilon_C - \epsilon_A)^2 + 4Z_A \beta^2}}, \qquad (1.4.62)$$

which may also be written:

$$E_{\text{cov}} = -2m|\beta|\sqrt{Z_A}\sqrt{\delta Q \left(2\frac{n_0}{m} - \delta Q\right)}, \qquad (1.4.63)$$

as a function of the anion–cation electron transfer $\delta Q = 2 - Q$. In a system with constant charges, E_{cov} is proportional to $\beta \sqrt{Z_A}$, as in metals. We will use these expressions in Chapters 2 and 3 to discuss the covalent contribution to the surface energy.

As an example, we give below the contributions to the cohesion energy of two simple insulators: MgO and NaCl, taken with reference to the ions, obtained by the self-consistent tight-binding method (Moukouri and Noguera, 1993).

- NaCl: $E_{\text{coh}} = 8.1$ eV; $E_{\text{coh}}^{\text{int}} \approx 0$ eV, $E_{\text{coh}}^{\text{cov}} = 5.1$ eV, $E_{\text{coh}}^{\text{M}} = 5$ eV, $E_{\text{coh}}^{\text{rep}} = -2$ eV.
- MgO: $E_{\text{coh}} = 40.4$ eV; $E_{\text{coh}}^{\text{int}} = 15.2$ eV, $E_{\text{coh}}^{\text{cov}} = 20.6$ eV, $E_{\text{coh}}^{\text{M}} = 12.7$ eV, $E_{\text{coh}}^{\text{rep}} = -8.1$ eV.

In NaCl, the relative importance of the Madelung term stresses the highly ionic character of this insulator. On the contrary, there exists an actual competition between electrostatic and covalent effects in MgO and in other oxides such as TiO_2 or SiO_2.

It should be noted that other partitionings of the total energy are also discussed in the literature. In particular, in methods which use pseudo-potentials in the LDA approximation, the energy is written in the following way (Yin and Cohen, 1982):

$$E = E_{\text{kin}} + E_{\text{ec}} + E_{\text{ee}} + E_{\text{xc}} + E_{\text{cc}}. \qquad (1.4.64)$$

The first term is the expectation value of the kinetic operator T on the ground state: $E_{\text{kin}} = \sum_{\vec{k}} \langle \Psi_{\vec{k}} | T | \Psi_{\vec{k}} \rangle$. The second and third terms are the electrostatic energies of interaction of the electrons with the cores V_i and with the electron density (Hartree term V_H). E_{xc} is the exchange and correlation energy, and finally, the core–core interactions are denoted E_{cc}. Leaving aside the exchange and correlation term, the correspondence between both partitionings is obtained by developing the Bloch wave functions $\Psi_{\vec{k}}$ on an orbital basis set $|j\rangle$ and writing separately the diagonal and non-diagonal matrix elements of the Hamiltonian on this basis set:

$$\sum_{\vec{k}} \langle \Psi_{\vec{k}} | T + \sum_i V_i + V_H | \Psi_{\vec{k}} \rangle = \sum_{\vec{k}j} a_{\vec{k}j} a_{\vec{k}j}^* \langle j | T + \sum_i V_i + V_H | j \rangle$$
$$+ \sum_{\vec{k}j \neq j'} a_{\vec{k}j} a_{\vec{k}j'}^* \langle j' | T + \sum_i V_i + V_H | j \rangle .$$

$$(1.4.65)$$

When associated with E_{cc}, the first term gives rise to the internal and Madelung energies, while the second term represents the covalent energy. We will use the summation made in (1.4.64) when discussing interfacial energies in Chapter 5.

1.5 Conclusion

Numerous studies have been devoted to bulk oxides, and their properties, both structural and electronic, are now well apprehended. At least, this statement applies to insulating oxides in which correlation effects do not induce a breakdown of the effective one-electron picture. In the remainder of this book, we will mainly focus on these 'simple' oxides, and we will describe how their properties are modified at surfaces, whether they are ideally clean and planar, or contain defects. With the purpose of interpreting these modifications in a unified theoretical framework, we have presented here the basis of a model which stresses the factors governing the electronic structure. These include: the relative position of the anions and cations in the periodic table, which fixes the energy difference between the outer levels of the neutral atoms $\epsilon_C^0 - \epsilon_A^0$; the strength of the Madelung potentials exerted on the ions, which depend upon the cation valency; the crystal structure and the inter-atomic distances; the values of the resonance integrals β and the coordination number, which determine to what extent the electrons are delocalized; and finally the optical dielectric constant, whose value is a measure of the strength of screening effects.

2

Atomic structure of surfaces

When a crystal is cut along some orientation, the atoms located in the few outer layers experience non-zero forces which are induced by the breaking of oxygen–cation bonds. Generally, they do not remain at the positions fixed by the three-dimensional lattice. Point or extended defects may result, as well as lattice distortions. This chapter analyses the structural features of oxide surfaces, which is also a useful step, before starting the discussion of the surface electronic properties. Yet, conceptually, this presentation is not fully satisfactory, because the structural and electronic degrees of freedom are coupled and both determine the ground state configuration. Despite a rich literature, the structural properties of oxide surfaces are not fully elucidated. It is often difficult to prepare stoichiometric and defect-free surfaces, and the characterization is hindered by charging effects and by an uncertainty about the actual crystal termination.

2.1 Preliminary remarks

We will make some preliminary remarks concerning the designation of the surfaces, their polar or non-polar character and their structural distortions – relaxation, rumpling and reconstruction.

Notations

A plane in a crystal, is identified by three integers (h, k, l), called the Miller indices, which are in the same ratio as $(1/x, 1/y, 1/z)$, the inverses of the coordinates of the intersections of this plane with the crystallographic axes (van Meerssche and Feneau-Dupont, 1977). Notations with four indexes $(h, k, -(h+k), l)$ are used in hexagonal structures, such as α-quartz, corundum α-alumina, or the wurtzite ZnO structure. If several planes are structurally equivalent, as the six (100), ($\bar{1}$00), (010), (0$\bar{1}$0), (001) and (00$\bar{1}$) planes in a rocksalt crystal, this collection is denoted by {100} in curly

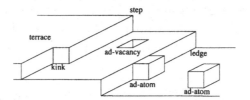

Fig. 2.1   Defects on surfaces.

brackets. Low-index surfaces are more compact and generally more stable than high-index surfaces, as developed later.

For strictly two-dimensional periodic structures, there exist five Bravais lattices (Kittel, 1990): square, rectangular, centred rectangular, hexagonal and oblique, and seventeen two-dimensional space groups. However, real surfaces rarely present a perfect ordering. Depending upon the conditions under which they are prepared – cutting, polishing, etc., – various defects may be found: steps with ledges and terraces, kinks, ad-atoms, ad-vacancies, etc., the most common of which are represented in Fig. 2.1.

Classification of surfaces

Since oxide surfaces involve at least two types of atom which bear charges of opposite sign, a classification into three families has been established on the basis of simple electrostatic criteria (Fig. 2.2) (Tasker, 1979a):

Type 1 surfaces   Each plane parallel to the surface is electrically neutral. The {100} and {110} faces of rocksalt crystals, in which the outer layers contain as many anions as cations, and the {110} faces of the cubic fluorite structure – found in UO_2 – with twice as many oxygens as cations, are type 1 surfaces.

Type 2 surfaces   In most oxide surfaces, the atomic layers are not neutral. However, the stacking sequence may be such that the total dipolar

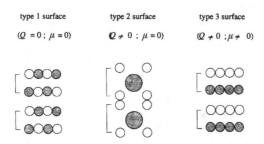

Fig. 2.2   Classification of insulating surfaces according to Tasker (1979a).

moment of the repeat unit vanishes. The rutile {110} faces, terminating with rows of bridging oxygens, belong to this family. The {111} faces of the fluorite structure are also of this type, provided that oxygens are in contact with vacuum.

Type 3 surfaces In the third family, the layers are not electrically neutral and the repeat unit bears a non-zero dipole moment. These are polar surfaces. The {111} faces of rocksalt crystals, the wurtzite {0001} faces, the fluorite {110} faces, etc., are polar. They deserve special attention due to their electric peculiarities. We postpone their analysis to Section 3.3.

Surface terminations

In complex oxide structures, it is generally not sufficient to know the surface orientation, because there exist several inequivalent bulk termi-nations. For example, strontium titanate, $SrTiO_3$, presents alternating layers of SrO and TiO_2 stoichiometries along the (001) direction. A (001) surface may thus have two distinct terminations (Fig. 2.3). Similarly, three distinct (110) cuts of a rutile crystal may be acheived, because of the six inequivalent layers of stoichiometries: O / 2TiO / O / O / 2TiO / O, along that direction. This also happens in α-alumina, on the $\left\{10\bar{1}2\right\}$ and {0001} surfaces. The actual termination of a surface is generally fixed by stability criteria and by the conditions of preparation, but it is rarely easy to determine it experimentally.

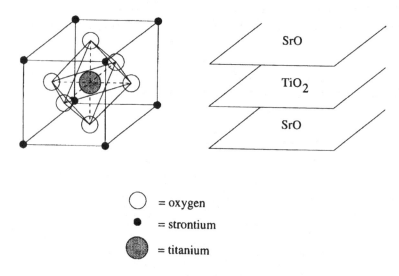

○ = oxygen
● = strontium
◉ = titanium

Fig. 2.3   Non-equivalent (001) layers in $SrTiO_3$.

Criteria for surface stability

Among all possible terminations, some are associated with a high surface energy. This is the case for polar terminations. Surface energies of type 1 or type 2 terminations are generally lower. Aside from this electrostatic consideration, an additional criterion of stability is provided by the number of oxygen–cation bonds which have been broken during the surface formation. Since energy is required to break these bonds, the most stable surfaces are those with the smallest number of broken bonds. They are also the most compact surfaces, on which the coordination number of the atoms is not greatly reduced. This criterion is well known in the surface physics of metals, although the prevailing microscopic contribution to the surface energy is different. For the rocksalt structure, the {100} cut, with one broken bond per surface atom, is the most stable, while the {110} and {211} orientations, with two and four broken bonds, respectively, have a higher surface energy and are difficult to obtain defect-free. Atomistic models have been used to calculate the surface energies of oxides crystallizing in various structures, such as rocksalt or corundum. In the latter case – which applies to e.g. α-alumina, Fe_2O_3, or Cr_2O_3 – on the basis of Madelung energy calculations, the models predict increasing surface energies along the series (Hartman, 1980; Bessières and Baro, 1973; Mackrodt *et al.*, 1987):

$$(10\bar{1}2) < (11\bar{2}0) < (10\bar{1}1) < (0001) < (10\bar{1}0)$$

In more covalent oxides, the principle of surface autocompensation (Biegelsen *et al.*, 1990a; 1990b) has been applied to predict the more stable surface terminations (Godin and LaFemina, 1994b). Starting from a description of bulk anion–cation bonds in which the cation valence n is equally shared between the Z first neighbours, this principle states that a surface is autocompensated if the charge involved in the cation dangling bonds is just sufficient to fill all the anion dangling bonds. When this condition is fulfilled, the surface termination presents an insulating character and a low energy.

Structural distortions

In Figs. 2.1 and 2.3, the atoms located in the outer layer, the ad-atoms or the ad-vacancies have been represented at positions defined by the bulk three-dimensional lattice. However, the bond-breaking process in the surface formation induces forces which push the atoms out of their bulk positions. When a two-dimensional periodicity is kept in the surface layers, the structural distortions are called: relaxations, rumplings or reconstructions (Fig. 2.4).

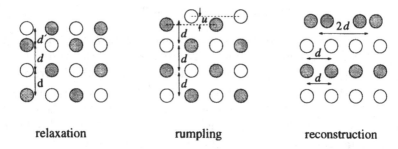

| relaxation | rumpling | reconstruction |

Fig. 2.4.   Relaxation, rumpling and reconstruction effects. In the first two cases, the surface cell is identical to the projection of the bulk three-dimensional cell but the inter-atomic distances between the surface and the underlying planes are modified. When a reconstruction takes place, the size of the surface cell changes.

Relaxation   A relaxation is associated with a change in the spacing between the surface top layer and the underlying plane. An inward relaxation corresponds to a spacing contraction. In the opposite case, an outward relaxation takes place. Relaxation effects may, and generally do, affect several layers.

Rumpling   Layers which contain at least two types of atom may also rumple, when the oxygen anions and the cations move in opposite vertical directions. The oxygens are generally displaced outwards and the cations inwards. This effect is not induced by specific properties of the oxygen atom. It is also found, for example, on ZnS $\{110\}$ surfaces. Rumpling and relaxation effects often occur simultaneously. The surface structure is then better described as the result of oxygen and cation relaxations having different strengths.

Reconstruction   A reconstruction is characterized by a change in the periodicity of the outer layer. For example, the (1×5) reconstruction of the rutile $\{100\}$ face indicates that the surface unit cell is in registry with the substrate cell but is five times expanded along the b direction. In the $(\sqrt{31} \times \sqrt{31})R9°$ reconstructed α-alumina $\{0001\}$ face, the lattice parameters are multiplied by $\sqrt{31}$ along the a and b directions and the surface cell is rotated by $9°$ with respect to the underlying layers. When a rotation occurs, it is more precise to give the rotation matrix which connects the unreconstructed and reconstructed lattices, but the simple notation given above is still very often used in the literature.

A lowering of the surface energy is induced by all the structural distortions which have just been described. It must be taken into account when comparing the relative stability of different faces of a given material.

2.2 Experimental and numerical studies

The techniques which are most extensively used to determine the surface atomic positions are electron diffraction techniques (LEED = Low-Energy Electron Diffraction and RHEED = Reflection High-Energy Electron Diffraction) – although a quantitative determination of the inter-plane distances requires heavy numerical simulations in parallel with these experiments – , ion backscattering, SEELFS (SEELFS = Surface Electron Energy Loss Fine Structure) and atomic force microscopy (AFM). The STM (Scanning Tunneling Microscope) can only be used for narrow gap oxides. When analysing a surface structure, it is useful to perform a precise chemical characterization of the outer layers simultaneously, by Auger and/or photoemission, to disantangle intrinsic surface processes from effects associated with stoichiometry defects produced during surface preparation. We summarize below the main results obtained on monocrystalline oxide surfaces, under well-controlled conditions, and refer the reader to the reviews of Henrich (1983; 1985) and Henrich and Cox (1994) for an extensive compilation.

Rocksalt surfaces

The {100} surface may be prepared by cleavage in most rocksalt oxides. The surface atoms are five-fold coordinated, while they occupy octahedral sites in the bulk. It is a type 1 surface with neutral layers containing an equal number of cations and oxygens. MgO{100} is the most thoroughly studied surface, by LEED – on ultra-high-vacuum or air cleaved samples (Kinniburgh, 1975; Welton-Cook and Berndt, 1982; Urano *et al.*, 1983; Blanchard *et al.*, 1991) – by RHEED (Maksym, 1985), by SEELFS (Goyhenex and Henry, 1992), by photoelectron diffraction (Varma *et al.*, 1994) and by medium-energy ion scattering (Zhou *et al.*, 1994). All results are consistent with an inward relaxation effect smaller than 1%. Ion backscattering work (Nakamatsu *et al.*, 1988), on the other hand, disagrees appreciably with this value, maybe because of the sample preparation conditions: mechanical polishing, high-temperature annealing after acid treatment. Different authors give different values for the rumpling, but all agree that it is very weak, typically smaller than a few hundredths of an ångstrom. The amplitude of the relaxation is qualitatively similar on the {100} faces of other rocksalt oxides. On NiO, an inward relaxation of 0.04 Å has been found (Kinniburgh and Walker, 1977), and an effect of the same order is reported on CoO (Felton *et al.*, 1979). On CaO, the structural distortions seem to be weaker than on MgO (Prutton *et al.*, 1979).

Other rocksalt surfaces, such as the {110} and the {111} faces, have higher surface energies and are thus less stable. Their surface atoms are respectively four-fold and three-fold coordinated. These faces often present steps or facets. Nevertheless, the polar {111} face has been studied on NiO (Floquet and Dufour, 1983) and MgO (Henrich, 1976; Onishi *et al.*, 1987; Henry and Poppa, 1990): it is a type 3 surface, terminated by either a cation or an oxygen layer. This face may be stabilized by the presence of impurities or when a thin oxide layer is grown on a substrate.

Rutile surfaces

Although rutile oxides do not cleave well, three low-index faces have been obtained. The {110} face is the most stable and, in the case of TiO_2, it can be prepared in a controlled and reproducible manner. Half of the surface titaniums are six-fold coordinated as in the bulk; the other half have lost one neighbouring oxygen. It is a type 2 surface when it is terminated by a layer of two-fold coordinated oxygens (bridging oxygens). The stacking sequence, from the surface, is then: O / 2TiO / O / O / 2TiO / O (Röhrer *et al.*, 1990; Maschhoff *et al.*, 1991). The {100} and {001} TiO_2 faces are respectively type 2 and type 1 surfaces, with the following repeat units: O / Ti / O / O / Ti / O for the {100} face and TiO_2 / TiO_2 / TiO_2 for the {001} face. Surface cations are five-fold and four-fold coordinated on the {100} and {001} faces, respectively. In agreement with the stability criterion relying on the number of broken bonds, the {100} face is more stable than the {001} one. On both surfaces, structural distortions, such as steps, facetings or reconstructions have been observed (Chung *et al.*, 1977; Tait and Kasowski, 1979; Henrich and Kurtz, 1981a; Wang *et al.*, 1994), to which we will return later.

Perovskite surfaces

The {100} face of a perovskite ABO_3 crystal is the most stable surface, although it cannot be prepared by cleavage. There exist two non-equivalent terminations with respective stoichiometries AO and BO_2 – for example, SrO and TiO_2 for $SrTiO_3$. In the surface layer, the transition metal cations B are five-fold coordinated, instead of six-fold in the bulk, and the cations A are surrounded by eight oxygens, instead of twelve in the bulk. The {100} surfaces of both $SrTiO_3$ and $BaTiO_3$ have been studied, but a quantitative surface structure determination has only been performed on the first compound: a contraction of about 10% on the SrO face and a weak expansion of the order of 2%, at the limit of the error bar, on the TiO_2 face, have been found by Bickel *et al.* (1989; 1990). Hikita *et al.* (1993), on the other hand, deduce an outward relaxation on both faces.

The $SrTiO_3${111} face may be prepared by sputtering and annealing. Depending upon the preparation mode, it remains atomically flat or develops facets (Lo and Somorjai, 1978). It is a type 3 polar surface with alternating SrO_3 and Ti layers.

Corundum surfaces

The α-Al_2O_3, α-Ti_2O_3, α-V_2O_3 and α-Fe_2O_3 oxides crystallize in the corundum structure, in which the cations occupy octahedral sites. The most stable face is the $\left\{10\bar{1}2\right\}$ – denoted $\{047\}$ by some authors – on which the cations are five-fold coordinated. The stacking sequence is: O / Al / O / Al / O. It is a type 2 surface. The $\{0001\}$ face may also be prepared by sputtering, in the case of α-Fe_2O_3, but it is distorted. In α-Al_2O_3, depending upon the preparation conditions, the $\{0001\}$ face is terminated either by oxygens or by aluminiums, or by layers of intermediate stoichiometries (Gautier *et al.*, 1991b). When aluminiums are located in the outermost layer, i.e. when the repeat unit is Al / 3O / Al / Al / 3O / Al, it is a type 2 surface. But when it is oxygen-terminated, with a sequence 3O / Al / Al / 3O / Al / Al, it is a polar surface. The strength of the structural distortions on these surfaces, which is predicted to be large by theorists, has not yet been established experimentally.

Wurtzite surfaces

In wurtzite oxides, the cations are not located in octahedral sites as in the previous structures. They are four-fold coordinated, which suggests a higher degree of covalency. ZnO is indeed at the border line between insulators and semi-conductors, with a gap width equal to 3.4 eV. The $\left\{10\bar{1}0\right\}$ and $\left\{11\bar{2}0\right\}$ faces have an equal number of anions and cations in each layer; they are non-polar, type 1 surfaces. The $\{0001\}$ and $\left\{000\bar{1}\right\}$ faces, on the other hand, are terminated by zinc and oxygen atoms, respectively: they are polar, type 3 faces. The $\left\{000\bar{1}\right\}$ and $\left\{11\bar{2}0\right\}$ do not present sizeable distortions, which seems strange at least for the polar face. On the $\{0001\}$ face, a strong relaxation, of the order of 0.4–0.5 Å has been found (Lubinsky *et al.*, 1976); on the $(10\bar{1}0)$ face, the relaxation is associated with a rumpling of the order of 0.4 Å (Duke *et al.*, 1978; Göpel *et al.*, 1982).

α-quartz surfaces

The α-quartz $(01\bar{1}0)$ face can be prepared by cutting the crystal perpendicular to the y-axis, and by subsequently performing a mechanical polishing (Bart *et al.*, 1992; 1994). It is a type 2 surface, presenting the following stacking sequence: Si / O / O / O / Si / O / O / O / Si. Depending upon the preparation mode, (1×1), (1×3), or (3×1) superstructures have been observed; the two latter have been assigned to other crystalline forms of SiO_2, such as β-quartz or tridymite. The α-quartz (0001) face has also been studied: it is a type 2 surface with a sequence: O / Si / O / O / Si / O / O / Si / O (Bart and Gautier, 1994).

Steps and kinks on oxide surfaces

Steps and kinks are among the most common defects on real surfaces. They involve atoms with a lower coordination than on low-index planar surfaces. They may be observed by LEED, ion scattering spectroscopy, atomic force microscopy or scanning tunneling microscopy for narrow gap oxides, and high-resolution transmission electron microscopy. This has been done, for example, on non-polar ZnO surfaces (Henzler, 1973; Cheng and Kung, 1981; Röhrer and Bonnell, 1991), rutile (Röhrer *et al.*, 1990), alumina (Hsu and Kim, 1991; Kim and Hsu, 1991), NiO (Smith *et al.*, 1986) and MgO(100) (Cowley, 1986; Kim *et al.*, 1993). Questions which have been discussed are the values of the step height, the orientation of the low-index faces which surround the steps on vicinal or facetted surfaces, etc. However, although high resolution transmission electron microscopy can image individual atoms, no quantitative determination of the bond lengths around the under-coordinated atoms has been achieved.

Numerical studies

Most numerical studies of the surface distortions have used atomistic approaches at various levels of sophistication. The calculated atomic displacements are very weak on $MgO\{100\}$ (Martin and Bilz, 1979; Tasker and Duffy, 1984; de Wette *et al.*, 1985; Mackrodt, 1988), $CaO\{100\}$ and $SrO\{100\}$ (Mackrodt, 1988). On $MgO\{110\}$ (Mackrodt, 1988) and on $UO_2\{111\}$ (Tasker, 1979b), the relaxation is found to be larger. On $SrTiO_3\{100\}$, Mackrodt (1989) has predicted a 7% rumpling. The calculations of Prade *et al.* (1993) agree with the experimental results of Bickel *et al.* (1989), which predict the existence of a 10% relaxation of the SrO face and nearly no relaxation on the TiO_2 face, provided that, on the latter, one introduces in the model the electron redistributions induced by the loss of one oxygen atom per TiO_6 octahedra. Mackrodt (1989) has calculated strong relaxations, a 16% expansion and a 24% contraction, on $ZrO_2\{110\}$ and $Li_2O\{111\}$ faces, respectively. In CeO_2, relaxation effects are found to increase in the surface series: (111), (110) and (310) (Sayle *et al.*, 1994). Interestingly, on the first surface, a rumpling with the cations moving outwards is predicted. Finally, simulations by Mackrodt *et al.* (1987), Tasker (1984), Mackrodt (1989), and Blonski and Garofalini (1993) predict strong distortions on α-alumina and other corundum surfaces.

Aside from these classical approaches, quantum calculations have also been performed, which simultaneously solve the electronic structure and optimize the geometry. The Hartree–Fock *ab initio* method has been applied to MgO (Causà *et al.*, 1986), Al_2O_3 (Causà and Pisani, 1989) and ZrO_2 (Orlando *et al.*, 1992). A tight-binding method has been used to study ZnO (Skinner and LaFemina, 1992), MgO, Al_2O_3, β-tridymite and β-cristobalite – two allotropic forms of SiO_2 – (La Femina and Duke, 1991; Godin and LaFemina, 1994a), and SnO_2 (Godin and LaFemina, 1993; 1994b). These studies have allowed relaxations with bond-length contraction in ionic oxides such as MgO to be distinguished from relaxations with bond rotations in more covalent oxides. Using a semi-empirical Hartree–Fock method, Goniakowski and Noguera (1995a) have studied the dependence of the rumpling and relaxation upon the surface orientation in MgO and TiO_2, and upon the oxide ionicity, in the series MgO(110), CaO(110), SrO(110) and BaO(110). SnO_2 (Manassidis *et al.*, 1995), TiO_2 (Vogtenhuber *et al.*, 1994; Ramamoorthy *et al.*, 1994) and α-Al_2O_3 surfaces (Manassidis *et al.*, 1993) have been modelled by the *ab initio* LDA method. This has confirmed the strong relaxations predicted by atomistic models in the last-named compound.

There have been several theoretical calculations of the relaxation of atoms around steps and kinks on oxide surfaces, most of them using atomistic approaches (Colbourn, *et al.*, 1983; Colbourn and Mackrodt, 1983; Tasker, 1984; Tasker and Duffy, 1984; Mackrodt and Tasker, 1985). These methods also yield the step energy and describe the interaction between steps as a function of their distance. To the author's knowledge, quantum methods have only been applied to stepped MgO(100) surfaces. They include the *ab initio* Hartree–Fock cluster method (Fowler and Tole, 1988; Pacchioni *et al.*, 1992a), the LDA method (Langel and Parrinello, 1994; Kantorovich *et al.*, 1995) and the semi-empirical Hartree–Fock method applied to vicinal surfaces (Goniakowski and Noguera, 1995c). A contraction of the lattice around the under-coordinated atoms is predicted by all authors, with an increasing strength as the atom coordination gets smaller. The inward relaxation is thus found to be stronger near outer edge atoms, corner atoms or kink atoms. A dilatation of the lattice close to the bottom edge of the steps is also found, leading to a smoothing of the irregularities. Qualitatively similar results are obtained around oxygen vacancies (Mackrodt and Stewart, 1977; Duffy *et al.*, 1984; Ramamoorthy *et al.*, 1994; Castanier and Noguera, 1995a).

2.3 Relaxations

A relaxation occurs when the inter-plane spacing between the surface top layer and the underlying plane differs from the bulk spacing. It may be associated with a surface rumpling when the layers contain more than one type of ion. Since the microscopic mechanisms underlying the two effects are different, we will consider them successively. In this section, we will mainly focus on relaxations which involve bond-length contractions, because their mechanism has been well studied in the literature. In some oxides a mechanism of relaxation via bond rotation, obeying a bond-length conservation principle has been invoked (LaFemina and Duke, 1991; Godin and LaFemina, 1994a), whose theoretical description remains to be developed.

In highly ionic compounds, the surface relaxation may be assigned to a competition between short-range repulsive forces and electrostatic attractions, produced by the bond breaking at the surface. In covalent materials, on the other hand, covalent rather than electrostatic forces are responsible for the surface relaxation. We will first develop these two limiting cases, before considering the more intricate problem of mixed iono–covalent materials.

Electrostatic mechanism

Before discussing the actual semi-infinite geometry which is relevant for surface effects, let us first consider a system in which all atoms are equally coordinated – a bulk material or some regular clusters, such as a square, a cube, etc., – and let us prove that the equilibrium inter-atomic distance R_0 is an increasing function of the coordination number Z.

Preliminary approach In Born's model (Section 1.1), the total energy per formula unit is the sum of an electrostatic term, proportional to the Madelung constant α, and of a contribution from the short-range repulsion between neighbouring atoms. We will express the latter in a Lennard–Jones form ($n \gg 1$):

$$E = \frac{-\alpha Q^2}{R} + \frac{Z A}{R^n} \; . \qquad (2.3.1)$$

The minimization of E with respect to R, yields R_0:

$$R_0 \propto \left(\frac{Z}{\alpha}\right)^{1/(n-1)} \; . \qquad (2.3.2)$$

When comparing various geometries, it turns out that the Madelung constant α increases with Z, less quickly than Z itself. For example, in the rocksalt structure: for a diatomic molecule, $Z = 1$ and $\alpha = 1$; for a square, $Z = 2$ and $\alpha = 1.29$; for a cube, $Z = 3$ and $\alpha = 1.46$; and in the bulk, $Z = 6$ and $\alpha = 1.75$. The equilibrium inter-atomic distance R_0 is thus an increasing function of Z. In NaCl, a simple estimate for these four geometries gives: $R_0 = 2.36$ Å, 2.51 Å, 2.60 Å and 2.79 Å, respectively.

This result supports the qualitative arguments, given in Section 1.1, which states that inter-atomic distances are equal to the sum of ionic radii, and that the latter increase with the coordination number (Shannon, 1976). It also supports the bond-length rules for adsorption processes which will be mentioned in Chapter 6 (Gutmann, 1978).

The compactness of the structure thus favours larger inter-atomic distances. A similar result is obtained whenever the attractive forces, responsible for the cohesion, vary with Z and R less quickly than the repulsive ones.

Improved derivation The derivation made above suggests that the bond lengths around surface atoms should be shorter than in the bulk, because of their reduced coordination. Yet, to prove the existence of an inward relaxation more rigorously, it is necessary to take into account the different environments of the bulk and surface atoms. To be specific, let us consider a rocksalt (100) face, and let us write down the energy terms which depend upon the distance d between the top layer and the underlying plane. The inter-plane spacing in the bulk is denoted d_0. The electrostatic interactions between planes which are an even number of times d_0 apart are repulsive, and attractive otherwise. Since the (100) planes are neutral, the absolute value of the interaction decreases exponentially with the distance: for two planes md_0 apart, we will write it: $\pm Y \exp(-qmd_0)$ ($Y > 0$, m integer). At short distances, the actual expression for the inter-plane interaction is a sum of exponentials involving the two-dimensional reciprocal lattice vectors. However, the calculations of the Madelung energy which make use of plane-by-plane summations show that, for neighbouring planes along dense orientations, the first term of the sum, $\exp(-qd_0)$, is already much smaller than 1. In MgO $\{100\}$ for example, $qd_0 \approx 3$–4 and $\exp(-qd_0) \approx 10^{-2}$. The d-dependent part of the total energy, per formula unit, thus reads:

$$E = -Y \exp(-qd) [1 - \exp(-qd_0) + \exp(-2qd_0) - ...] + \frac{A}{d^n}. \quad (2.3.3)$$

The first term is the electrostatic energy and the second one accounts for pair repulsions at short distances. Similarly, in the bulk, the d_0-dependent part of the energy is equal to:

$$E = -2Y \exp(-qd_0) [1 - \exp(-qd_0) + \exp(-2qd_0) - ...] + \frac{2A}{d_0^n}. \quad (2.3.4)$$

The minimizations of E with respect to d_0, in (2.3.4), and with respect to d, in (2.3.3) yield an implicit relationship between d and d_0:

$$\left(\frac{d}{d_0}\right)^{n+1} = \frac{\exp[q(d_0 - d)]}{[1 + \exp(-qd_0)]}. \quad (2.3.5)$$

At the lowest order of perturbation, the relaxation is equal to:

$$\frac{d - d_0}{d_0} \approx -\frac{\exp(-qd_0)}{(n + 1 - qd_0)} .$$

(2.3.6)

In a highly ionic system, $d - d_0$ is thus negative ($n + 1$ is generally larger than qd_0), and an inward relaxation takes place. Its strength is related to the value of $Y \exp(-qd_0)$, i.e. roughly to the difference in the Madelung potential acting on bulk and surface atoms. The relaxation is thus the weakest on the dense surfaces which have the largest Madelung constant. If the distance d_1 between the second and third layers is also varied in the model, at the lowest order of perturbation, d_1 is given by:

$$\frac{d_1 - d_0}{d_0} \approx \frac{\exp(-2qd_0)(n + 1 - 2qd_0)}{(n + 1 - qd_0)^2} .$$

(2.3.7)

The modifications of the inter-plane spacings decrease exponentially inside the solid. Depending upon the sign of $(n + 1 - 2qd_0)$, the attenuation is either monotonic or oscillatory.

The mechanism described above explains the relaxation effects found in atomistic descriptions of ionic surface structures. It mainly involves a competition between electrostatic and short-range repulsion forces.

Covalent mechanism

In many materials, the cohesion does not result from electrostatic interactions. In metals, for example, the electron delocalization due to the orbital hybridization is the driving force for cohesion. Metal surfaces have been studied for a long period of time. Most of them also present relaxation effects which have the following characteristics (Desjonquères and Spanjaard, 1993):

- an *inward* relaxation generally occurs,
- the inter-plane spacing contraction is larger on less dense surfaces,
- several spacings may be perturbed. When this occurs, the relaxation decreases in an oscillatory way inside the solid. The attenuation length is shorter when the surface compactness is larger,
- when adsorbates are deposited on a relaxed surface, the relaxation generally disappears.

It is interesting to study the mechanism underlying these effects, in detail, and make a comparison with the electrostatic arguments developed above.

Preliminary approach   As in the previous analysis, we will first assume that all atoms have the same coordination number Z. In metals, the cohesion energy has two main contributions, the covalent energy and the atom–atom repulsion at short distances. The dependence of the covalent term on the inter-atomic distance R and on Z may be derived by an analytical tight-binding approach. The density of states $N(E)$ is modelled by a gaussian function, whose centre of gravity is equal to the mean atomic energy ϵ_0 of the metallic atom outer levels, and whose width is equal to the square root of $\mu = Z\beta^2$, with β the first-neighbour resonance integral (Equation (1.4.40) in Chapter 1). The density of states may thus be written in the following form:

$$N(E) = \sqrt{\frac{1}{2\pi\mu}} \exp\left(-\frac{(E - \epsilon_0)^2}{2\mu}\right) , \qquad (2.3.8)$$

and the covalent energy is equal to:

$$E_{\text{cov}} = 2 \int_{-\infty}^{E_F} (E - \epsilon_0) N(E) dE , \qquad (2.3.9)$$

i.e.:

$$E_{\text{cov}} = -\sqrt{\frac{2\mu}{\pi}} \exp\left(-\frac{(E_F - \epsilon_0)^2}{2\mu}\right) . \qquad (2.3.10)$$

On the other hand, the number of electrons N in the outer atomic orbitals determines the Fermi energy E_F:

$$N = 2 \int_{-\infty}^{E_F} N(E) dE . \qquad (2.3.11)$$

N depends only on the dimensionless parameter $(E_F - \epsilon_0)/\sqrt{2\mu}$. This proves that E_{cov} is the product of $\sqrt{\mu}$ by a dimensionless factor $-B$ $(B > 0)$, which is fixed by N:

$$E_{\text{cov}} = -B|\beta|\sqrt{Z} . \qquad (2.3.12)$$

The resonance integrals β are decreasing functions of the inter-atomic distance R: $\beta = \beta_0 / R^m$ and the short-range repulsion term may be chosen in a Lennard–Jones form. The contributions to the total energy per atom thus vary with R and Z as:

$$E_{\text{cov}} = -\frac{B_0 \sqrt{Z}}{R^m} , \qquad (2.3.13)$$

and

$$E_{\text{rep}} = \frac{Z A}{R^n} . \qquad (2.3.14)$$

The minimization of $E = E_{\text{cov}} + E_{\text{rep}}$ with respect to R yields the equilibrium inter-atomic distance R_0:

$$R_0 = \left(\frac{nA\sqrt{Z}}{mB_0}\right)^{1/(n-m)} , \qquad (2.3.15)$$

while the stability of the structure requires that $(n - m) > 0$. According to (2.3.15), R_0 is an increasing function of the coordination number Z and one can prove that this result does not depend upon the R-laws chosen for the resonance integrals and the short-range repulsion. This approach was used to explain the observed and/or calculated distance contractions in small metallic clusters (Joyes, 1990).

The conclusion concerning the inter-atomic distance is thus similar in ionic and in metallic systems, despite the difference in the cohesive forces. The variations of the equilibrium distance with Z come from the fact that the attractive forces vary with Z and R less rapidly than the repulsive forces.

A better description of metal surfaces   A better description of the surface relaxation is obtained by explicitly writing down the local densities of states in the top layer (S) and in the sub-surface plane (SS). The surface atoms have Z_c missing neighbours, $Z - 2Z_c$ bonds with surface atoms and Z_c bonds with atoms in the sub-surface plane. In the surface plane, the bond length is assumed to be unchanged and the resonance integrals are denoted by β_0. The Z_c inter-plane bonds are (possibly) shortened, and, along them, the resonance integrals are written as β. The covalent energy associated with surface atoms thus reads:

$$E^S_{cov} = -B\sqrt{(Z - 2Z_c)\beta_0^2 + Z_c\beta^2} \ . \tag{2.3.16}$$

The atoms in the sub-surface plane have $Z - Z_c$ bonds with unchanged lengths d_0 ($Z - 2Z_c$ in their own plane and Z_c with the atoms located in the plane below) and Z_c bonds (possibly) shortened with atoms in the surface plane. Their covalent energy is thus equal to:

$$E^{SS}_{cov} = -B\sqrt{(Z - Z_c)\beta_0^2 + Z_c\beta^2} \ . \tag{2.3.17}$$

The minimization of the total energy $E = E^S_{cov} + E^{SS}_{cov} + E_{rep}$ with respect to d leads to an implicit equation for d (Desjonquères and Spanjaard, 1993). The distance d is found to be smaller than d_0, which can be checked straightforwardly by performing a first-order expansion with respect to Z_c/Z. The contraction of the inter-plane bond lengths is a sign of an inward relaxation. The relaxation of the other layers has to be solved numerically.

Relaxation mechanism in mixed iono–covalent systems

The oxygen–cation bonds in insulating oxides have a mixed ionic and covalent character. It is thus tempting to merely interpolate the previous results and conclude the existence of an inward relaxation on their surfaces.

However, great care must be taken in doing so, as revealed by a study of small iono–covalent clusters (Moukouri and Noguera, 1992; 1993).

In oxides, the total energy has three main contributions associated with short-range repulsion, Madelung and covalent interactions:

$$E = \frac{ZA}{R^n} - \frac{\alpha Q^2}{R} + E_{\text{cov}} \,. \tag{2.3.18}$$

As demonstrated in Chapter 1, Equation (1.4.63), using the alternating lattice approach and the delta function approximation for $M(F)$, the covalent energy E_{cov} can be expressed in the following way:

$$E_{\text{cov}} = -B|\beta|\sqrt{Z}\sqrt{\left(2\frac{n_0}{m} - \delta Q\right)\delta Q} \,, \tag{2.3.19}$$

as a function of the difference δQ between the oxygen formal charge $Q_{\text{max}} = 2$ and its actual charge $|Q_0|$ in the system ($\delta Q > 0$). With the notations used in (1.4.48), δQ reads:

$$\delta Q = \frac{n_0}{m}\left(1 - \frac{\epsilon_C - \epsilon_A}{\sqrt{(\epsilon_C - \epsilon_A)^2 + 4Z\beta^2}}\right) \,. \tag{2.3.20}$$

The dependence of the equilibrium inter-atomic distance R_0 on the co-ordination number Z is obtained by minimizing E with respect to R. At variance with the discussion made for ionic systems, one has to be careful that, in some oxides, the charge may vary significantly with Z. For example, the ionic charge which accounts for the dipole moment measured in the MgO diatomic molecule is close to 0.6, while in the bulk all authors agree that it is larger than 1. Q thus varies at least by a factor of 2 when Z changes from 1 to 6. Under some circumstances, this effect may invert the tendency of bond contraction around under-coordinated atoms. Let us, for example, consider the competition between the repulsion and Madelung terms. The minimization of $E_M + E_{\text{rep}}$ with respect to R yields:

$$R_0 \propto \left(\frac{Z}{Q^2\alpha}\right)^c \,, \tag{2.3.21}$$

with $c = 1/(n-1) > 0$. In some systems such as MgO and CaO, in which the Z dependence of the charge Q is strong, the denominator increases with Z more quickly than the numerator, thus inducing a bond-length *expansion* when Z decreases. On the other hand, the charge variation has nearly no effect on the competition between the covalent energy and the short-range repulsion. Three cases may thus occur:

- E_{cov} and E_M vary with Z more quickly than E_{rep}. A bond-length expansion takes place when Z decreases.
- E_{cov} and E_M vary with Z less quickly than the repulsion term. A bond-length contraction takes place when Z decreases.
- One term varies more quickly than the repulsion term while the other varies less quickly; the qualitative trend is fixed by the larger term in absolute value.

In the NaCl, CaO and MgO clusters, the first case never happens. The second case is exemplified by NaCl clusters, in which the charge variation is weak. The third behaviour characterizes oxide clusters such as MgO and CaO, in which the bond has a non-negligible part of covalent character and the ionic charge is very dependent on the cluster size. A contraction effect is found in numerical calculations and when comparing the experimental bond lengths in the diatomic molecule and in the crystal. This shows – and this point is important but not well recognized – that the covalent energy is, in absolute value, larger than the Madelung energy in these oxides.

In iono–covalent systems, the competition between covalent and ionic processes has thus a subtle role on structural phenomena. As far as non-polar surfaces are concerned, very few numerical approaches have quantitatively determined the covalent and electrostatic contributions to the surface energy. It is thus difficult to conclude on their relative importance in this or that oxide. In most non-polar surfaces, as we will see in the next chapter, the surface charges are very close to the bulk ones. The difficulty met in small clusters is thus absent. A systematic inward relaxation has indeed been found in all models of the relaxation in rocksalt oxides, rutile, quartz, α-alumina, etc. When several surfaces of a given material were studied, it was found that relaxation effects are stronger when the number of broken bonds at the surface is larger. A theoretical model of relaxation was proposed, on the basis of numerical results obtained by the semi-empirical Hartree–Fock method (Goniakowski and Noguera, 1995a), which showed that, in rocksalt oxides and rutile, the relaxation is mainly driven by the competition between the covalent and the short-range repulsion terms. These two terms alone were thus kept in the model which, as a consequence, is mathematically similar to the model presented for metal surfaces. Nevertheless, the covalent energy was written in a form suited for oxides, (2.3.19), proportional to $\beta\sqrt{Z}$ when the oxygen–cation charge transfer δQ remains constant.

A similar model accounts for the bond-length contraction around atoms located at step edge or on a kink (Goniakowski and Noguera, 1995c). It assumes that the modifications of a given bond length is independent of the neighbouring bonds. It thus considers only the part of the energy which depends on a given bond length R between two atoms of respective coordination numbers Z_1 and Z_2. The $Z_1 - 1$ and $Z_2 - 1$ other bond lengths are assumed to be equal to the bulk inter-atomic distance R_0:

$$E = -B\sqrt{(Z_1 - 1)\beta_0^2 + \beta^2} - B\sqrt{(Z_2 - 1)\beta_0^2 + \beta^2} + 2\frac{A}{R^n} . \qquad (2.3.22)$$

The last term is the short-range repulsion. The two first ones give the covalent energy. For resonance integrals varying as $1/R^m$, the equilibrium bond length is equal to:

$$\left(\frac{R_0}{R}\right)^{n-2m} = \frac{\sqrt{6}}{2}\left(\frac{1}{\sqrt{Z_1}} + \frac{1}{\sqrt{Z_2}}\right) . \qquad (2.3.23)$$

This equation proves that atoms of lower coordination numbers tend to come closer to their first neighbours and that the bond length decreases as Z_2 decreases, at constant Z_1. In MgO(100) with $m = 2$ and $n = 5.655$, it predicts, for example, an 8.6% bond contraction between an upper edge atom ($Z_1 = 4$) and a terrace atom ($Z_2 = 5$). The model applies as well to relaxation effects on planar surfaces. It predicts a 2.8% relaxation of MgO(100) terraces ($Z_1 = 5$, $Z_2 = 6$), in agreement with most numerical calculations.

These models of relaxation suggest that a bond-length contraction should also occur around bulk and surface vacancies. However, they have to be modified to account for the charge redistributions which take place around such defects.

2.4 Rumpling

The surface rumpling is induced by the existence of forces of different strength acting on the anions and on the cations. It occurs on all surfaces which involve more than one type of atoms, along with the relaxation. With respect to the undistorted planar geometry, on a rumpled surface the anions are generally located more outwards than the cations. One exception is the cation outward displacements predicted by Sayle *et al.* (1994) on CeO$_2$(111). The rumpling amplitude is weak on dense rocksalt oxide surfaces – of the order of 0.05 Å on MgO{100} – and much stronger in more covalent systems – about 0.4–0.5 Å on the non-polar ZnO surfaces. Two mechanisms have been proposed in the literature to account for it. One relies on the existence of different electrostatic forces on the anions and cations, while the second invokes covalent forces. We will consider them successively.

Electrostatic mechanism

On a {100} face of a rocksalt crystal, Verwey (1946) has shown that the electrostatic forces acting on the surface cations and anions, and especially those arising from the atomic polarization, induce different displacements of the two species. Indeed, the ion polarizability is the only electrostatic parameter which can break the anion–cation symmetry on this surface. Verwey introduced the anion polarizability α' in a classical Born's model, and neglected the cation one, which is known to be much smaller due to the relative sizes of the species. The polarization of the bulk atoms, which is not relevant for surface effects, was neglected. Sawada and Nakamura (1979) have extended the model to finite temperatures, and have shown that the rumpling strength is an increasing function of temperature.

To derive the sign of the effect, we will first analyse the forces acting on the surface atoms in the *undistorted* geometry. We will find that both species are pulled inside and that the total force acting on the anions is weaker than on the cations.

The ions which belong to the sub-surface planes exert two types of forces on the outer layer: short-range repulsive forces, which push the surface ions outwards, and electrostatic (Madelung) forces which pull them inwards. Due to the symmetry of the lattice, these forces are equal on both species. Since the short-range forces are weaker than the Madelung ones, an inward relaxation of the surface layer results, as discussed in the previous section.

The surface anions are polarized by the electric field $\vec{\mathscr{E}}$ which acts on them, and a dipole moment $\vec{\mu} = \alpha'\vec{\mathscr{E}}$ is induced. The contribution of the surface charges to $\vec{\mathscr{E}}$ vanishes by symmetry when the atoms are coplanar. $\vec{\mathscr{E}}$ is thus due to the underlying layers. It is normal to the surface and, at the position of an anion, it is oriented outwards. The induced dipole moments $\vec{\mu}$ have the same orientation on each surface anion. They are submitted to a force proportional to the z component of the electric field gradient. To estimate the contribution of the surface layer to $\partial\mathscr{E}_z/\partial z$, one writes down the value of the electrostatic potential at a distance z above an anion:

$$V = \sum_i \frac{Q_i}{(R_i^2 + z^2)^{\frac{1}{2}}} . \tag{2.4.1}$$

The normal component of the electric field is equal to:

$$\mathscr{E}_z = -\frac{\partial V}{\partial z} = \sum_i \frac{Q_i z}{(R_i^2 + z^2)^{\frac{3}{2}}} , \tag{2.4.2}$$

and its gradient at $z = 0$:

$$\frac{\partial\mathscr{E}_z}{\partial z} = \sum_i \frac{Q_i}{R_i^3} . \tag{2.4.3}$$

The contributions of the underlying layers may be estimated by similar methods. They are much smaller than the surface contribution. The resulting electric field gradient is thus positive and the force exerted on the anions pushes them towards vacuum.

The total force, which acts on the surface atoms, pulls them inwards but, due to polarization effects, it is weaker on anions than on cations. This explains the origin and the sign of the surface rumpling.

More quantitatively, let us consider a rumpled rocksalt $\{100\}$ surface, in which the anions and cations in the outer layer are respectively located at d_1 and d_2 from the sub-surface plane. To be specific, we will assume that d_1 is larger than d_2, and we will check at the end of the calculation that this inequality is indeed fulfilled. The energy per formula unit contains several terms which are functions of d_1 and d_2.

The short-range repulsion energy, written in a Lennard–Jones form, reads:

$$\Delta E_1 = \frac{A}{d_1^n} + \frac{A}{d_2^n} + \frac{4A}{\left[d_0^2 + (d_1 - d_2)^2\right]^{n/2}}, \tag{2.4.4}$$

with d_0 the bulk inter-atomic distance. The two first terms give the interaction between the outer plane and the underlying plane. The third term contains the interactions between atoms located in the surface layer.

The electrostatic interaction between the surface charges and the underlying planes is equal to:

$$\Delta E_2 = -\frac{Y \exp(-qd_1)}{1 + \exp(-qd_0)} - \frac{Y \exp(-qd_2)}{1 + \exp(-qd_0)}, \tag{2.4.5}$$

with the same approximation for the inter-plane electrostatic interaction as in Section 2.3. The electric field due to the substrate, which acts on an anion reads:

$$\mathcal{E}_{2z} = \frac{Y q \exp(-qd_1)}{Q\left[1 + \exp(-qd_0)\right]}. \tag{2.4.6}$$

Inside the surface layer, to lowest order in $(d_1 - d_2)$, only anion–cation electrostatic interactions give a contribution to the energy:

$$\Delta E_3 = \frac{Q^2 C_1 (d_1 - d_2)^2}{d_0^3}. \tag{2.4.7}$$

The factor C_1/d_0^3 results from the two-dimensional summation of $1/R_i^3$ interactions between a given ion and all charges of opposite type. ΔE_3 is a positive quantity, because it is the energy required to pull apart charges of opposite signs. The electric field exerted by the surface layer on a surface anion is equal to:

$$\mathscr{E}_{3z} = \frac{2QC_1(d_1 - d_2)}{d_0^3} .$$ (2.4.8)

In response to the electric fields $\mathscr{E}_{2z}$ and $\mathscr{E}_{3z}$, a dipole moment μ is induced on the surface anions. Actually, the total electric field is equal to $\mathscr{E}_{2z} + \mathscr{E}_{3z} + \mathscr{E}_{4z}$, taking into account the contribution $\mathscr{E}_{4z}$ of the surface dipoles:

$$\mathscr{E}_{4z} = -\sum_i \frac{\mu}{R_i^3} = -\frac{C_2\mu}{d_0^3} .$$ (2.4.9)

The factor C_2/d_0^3 comes from the summation of the $1/R_i^3$ interactions between a given dipole moment and all the other anion dipoles (C_2 is thus different from C_1). A self-consistent equation relates the induced moment μ to the total electric field $\mathscr{E}_z$:

$$\mu = \alpha'(\mathscr{E}_{2z} + \mathscr{E}_{3z} + \mathscr{E}_{4z}) .$$ (2.4.10)

From (2.4.6) to (2.4.10), one derives the value of $\mathscr{E}_z$:

$$\mathscr{E}_z = \frac{1}{(1 + \alpha'C_2/d_0^3)} \left(\frac{Yq\exp(-qd_1)}{Q[1 + \exp(-qd_0)]} + \frac{2QC_1(d_1 - d_2)}{d_0^3} \right) .$$ (2.4.11)

Since both $\mathscr{E}_{2z}$ and $\mathscr{E}_{3z}$ are positive and since $\mathscr{E}_{4z}$ is negative but weaker, the z-component of the total electric field $\mathscr{E}_z$ is positive, and so is μ. The dipole formation energy is equal to:

$$\Delta E_4 = -\frac{\alpha'\mathscr{E}_z^2}{2} .$$ (2.4.12)

The equilibrium distances d_1 and d_2 are obtained by minimization of the total energy. In order to display more clearly the rumpling effect, it is convenient to introduce the parameters $\bar{d}$ and δ ($d_1 = \bar{d} + \delta$ and $d_2 = \bar{d} - \delta$). In the absence of anion polarizability ($\alpha' = 0$), the rumpling amplitude 2δ vanishes and $\bar{d}$ has the value already obtained in the study of relaxation effects (Section 2.3). The two first terms in the expansion of the total energy with respect to δ are of zeroth and second order:

$$\Delta E = \Delta E_1 + \Delta E_2 + \Delta E_3 = \Delta E_0 + x\delta^2 ,$$ (2.4.13)

with x positive. The minimum energy corresponds to $\delta = 0$.

When α' is non-zero, the expansion of ΔE contains a linear term provided by ΔE_4:

$$\Delta E = \Delta E_0 + y\delta + x'\delta^2 ,$$ (2.4.14)

with y equal to:

$$y = -\frac{\alpha' Y q \exp\left(-q\bar{d}\right)}{\left(1 + \alpha' C_2/d_0^3\right) Q \left[1 + \exp\left(-qd_0\right)\right]}$$
$$\times \left(\frac{8QC_1}{d_0^3} - \frac{Y q^2 \exp\left(-q\bar{d}\right)}{Q \left[1 + \exp\left(-qd_0\right)\right]}\right). \qquad (2.4.15)$$

In the last factor, the first term comes from the contribution ΔE_3 of the surface layer, and the second term is due to the underlying layers (ΔE_2 term). The former being the larger, y is negative. On the other hand, x' is positive, as in the case where $\alpha' = 0$. The rumpling amplitude at equilibrium $2\delta = d_1 - d_2 = -y/x'$ is thus positive. This corresponds to a relative displacement of the anions towards vacuum ($d_1 > d_2$). It gives rise to a geometric dipole moment whose sign is opposite to that of $\mu = \alpha' \mathscr{E}_z$.

Verwey and, after him, the authors who have introduced the polarization effects in the Born model, have shown that the rumpling amplitude correlates well with the difference in the anion and cation ionic radii. For example, along the series LiF, LiCl, LiBr and LiI, Verwey predicts rumpling amplitudes equal to 0.165 Å, 0.341 Å, 0.411 Å and 0.546 Å, respectively, while the cation ionic radii are equal to 1.36 Å, 1.81 Å, 1.95 Å and 2.16 Å, respectively. The geometric dipole moment balances a large part of the anion dipole moment μ. For example, in LiF, Verwey finds $\mu = 0.59$ Debye, and a geometric dipole moment equal to -0.79 Debye. Many calculations have been performed to determine the rumpling amplitude on MgO{100}, either using atomistic approaches or taking into account the electronic structure. In these latter, 2δ is found to be smaller than 0.05 Å, while shell models give more scattered values, ranging from 0.04 Å to 0.23 Å. A review of these calculations is given by LaFemina and Duke (1991).

Covalent mechanisms

The rumpling observed on compound semi-conductor surfaces is generally much stronger than on rocksalt oxide surfaces, despite a smaller difference in the anion and cation polarizabilities. This suggests that another rumpling mechanism is at work in these materials. After a long controversy, the electronic origin of the rumpling on the non-polar {110} face of blende compounds has been settled (Lannoo and Friedel, 1991; LaFemina, 1992). One argument relies on the fact that, in these materials, it is easier to bend a bond than to change its length, because the radial elastic constant is higher than the angular ones. The structural distortions thus

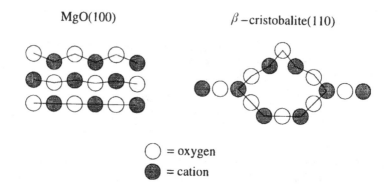

MgO(100) β –cristobalite(110)

○ = oxygen
● = cation

Fig. 2.5. Rumpling distortions on an MgO{100} and β-cristobalite{110} face (side view). One should note the bond contraction in the first oxide and the bond rotation in the second.

obey a bond-length conservation principle, whenever possible (Fig. 2.5). In addition, in the absence of surface distortions, two bands of surface states located in the bulk gap are predicted, which are associated with the cation and anion dangling bonds. However, no intrinsic surface states are observed experimentally. When gap states were found, they were assigned to defect states.

According to an argument first proposed by Chadi (1979), it was recognized that a surface rumpling with the anions outwards increases the hybridization of the anion dangling bond with the orbitals of the underlying atoms. This lowers its energy and pushes it into the valence band. A symmetric effect is at work on the cation dangling bond, which becomes degenerate with the conduction band states. The gap width is thus equal to its bulk value, in agreement with experimental measurements. The stabilization of the filled states in the rumpled geometry provides an energy lowering which is the actual origin of the rumpling effect. Typical values of the rumpling amplitude are 0.6 to 0.9 Å for semi-conductor compounds from GaP to InSb (Lannoo and Friedel, 1991). In covalent oxides, calculations predict rumplings of the order of 0.5 Å for the $\{10\bar{1}0\}$ and $\{11\bar{2}0\}$ ZnO faces, the β-tridymite $\{10\bar{1}0\}$ faces and the β-cristobalite SiO_2 {110} face (LaFemina and Duke, 1991).

The covalent mechanism described above requires specific features of the electronic structure and of the cohesion properties. Consequently, it does not apply to all oxide surfaces. However, there exist other covalent forces, present in most oxides, which can also lead to a surface rumpling. The direct oxygen–oxygen or cation–cation electron delocalization processes have indeed different strengths, because of the different sizes of the two species (Goniakowski and Noguera, 1995a). Since anions are generally larger than cations, the oxygen–oxygen resonance integrals are larger than the cation–cation ones, even if the inter-atomic distances are equal. This is also confirmed by the fact that such delocalization processes have to be considered when one wishes to reproduce correctly the valence band width in oxides (Chapter 1). Depending upon the sign of the anion–anion covalent energy, an expansion or a contraction of the corresponding bond lengths can lead to an energy lowering. For example, on $MgO\{100\}$, the oxygen–oxygen covalent energy is positive, so that an expansion of the oxygen–oxygen bond length stabilizes the system. This is indeed what occurs when the surface is rumpled with the oxygens outwards. In other oxides, the reverse may occur, as proposed for $TiO_2(001)$. This mechanism may thus act in phase with the electrostatic mechanism or compete with it. No quantitative estimate of their relative strengths has been done up to now, and there is a lack of systematic experimental results with which to compare.

2.5 Non-stoichiometric reconstructions

Contrary to metal and semi-conductor surfaces which present many intrinsic reconstructions, most observed reconstructions on oxide surfaces may be assigned either to an ordering of oxygen vacancies, characteristic of a surface sub-stoichiometry, or to a manifestation, at the surface, of a structural change in the bulk of the material.

This latter case is exemplified by the doubling of the two-dimensional surface cell observed on the (001) face of $SrTiO_3$, which is associated with the bulk anti-ferrodisplacive structural phase transition at 105 K. We will come back to this phase transition in Chapter 4 when discussing soft phonons.

Another example is provided by the (1×2) surface reconstruction observed in α-Fe_2O_3 $(10\bar{1}2)$ after heating at 1173 K. It was first assigned to an ordering of oxygen vacancies along distorted rows of atoms, yielding a doubling of the unit cell in one direction (Lad and Henrich, 1988). It was reconsidered recently in the light of LEED and Auger results. The change from the (1×1) to the (1×2) LEED patterns was found to take place at a temperature of the order of 723 K, and to be associated with a reduction of the O/Fe ratio of the order of 17%. It was related to a bulk phase transition between α-Fe_2O_3 and Fe_3O_4. This phase transition could be thought of as being initiated at the surface of the sample and progressively affecting more and more layers as the temperature increases (Pollak, 1995). In the (0001) face of α-Fe_2O_3, the stoichiometry is also very sensitive to thermal treatment. Depending upon the conditions of preparation, a (111) layer of magnetite Fe_3O_4 or islands of wüstite $Fe_{1-x}O$

have been observed at the surface (Lad and Henrich, 1988; Condon *et al.*, 1994; Barbieri *et al.*, 1994).

On the $(01\bar{1}0)$ surface of α-quartz, (3×1) and (1×3) LEED patterns have been observed after annealing at 1173 K. They were correlated either to the bulk $\alpha \rightarrow \beta$ transition which occurs in SiO_2 at 846 K or to a tridymite phase present on the surface (Bart *et al.*, 1992; 1994). On the (0001) surface, a reconstruction was also observed by LEED, after heating in air above 873 K. A preliminary interpretation in term of the $\alpha \rightarrow \beta$ phase transition was proposed (Bart and Gautier, 1994).

Several oxides display reconstructed LEED patterns which may be assigned to the presence and the ordering of oxygen vacancies in the outer layers. Leaving aside the polar surfaces which will be considered in the next chapter, this is the case in the rutile $TiO_2(100)$ face, which presents (1×3) , (1×5) or (1×7) superstructures, depending upon the annealing temperature: $T \approx 873$ K, $T \approx 1073$ K and $T \approx 1473$ K, respectively (Zschack *et al.*, 1992). Relying on an older study, in which the surface stoichiometry was controlled by Auger spectrometry (Chung *et al.*, 1977), the authors proposed two models of oxygen vacancy ordering on the surface: the missing row model and the microfacet model. In the latter, for example, the surface presents streaks, parallel to the surface [001] direction. The streak slopes are {110} facets, the most stable in the rutile structure. They are three planes deep for the (1×3) reconstruction, five planes deep for (1×5), etc. The microfacet structure was confirmed by a recent study and the atomic configuration around steps on the microfacetted surface was analysed (Hardman *et al.*, 1993; Murray *et al.*, 1994a; 1994b). The (110) face of rutile TiO_2 also displays reconstructions upon reduction, involving an ordering of the bridging oxygen rows (Sander and Engel, 1994; Onishi and Iwasawa, 1994), and the same is true for $SnO_2(110)$ (Cox *et al.*, 1989).

The $SrTiO_3(001)$ surface, terminated by a TiO_2 plane, was analysed by STM (Matsumoto *et al.*, 1992; Liang and Bonnell, 1993). After annealing at $T \approx 1200$ K, it presents a (2×2) superstructure, interpreted as an oxygen vacancy ordering along both directions of the two-dimensional mesh. A $(\sqrt{5} \times \sqrt{5})R26.6°$ reconstruction was also found on this surface termination, by recording an STM image at a bias such that tunneling takes place through the gap state induced by the oxygen vacancies (Tanaka *et al.*, 1994).

The under-stoichiometric α-alumina surfaces present rich reconstructions associated with oxygen vacancies. After annealing at increasing temperatures, a succession of reconstruction patterns was observed on the (0001) surface: (1×1), $(\sqrt{3} \times \sqrt{3})R30°$, $(2\sqrt{3} \times 2\sqrt{3})R30°$, $(3\sqrt{3} \times 3\sqrt{3})R30°$ and $(\sqrt{31} \times \sqrt{31})R \pm 9°$ (Chang, 1968; French and Somorjai, 1970; Baik *et al.*, 1985; Pham Van *et al.*, 1990; Gautier *et al.*, 1991b; 1994). Each of these reconstructions is associated with a different oxygen content, the density of vacancy increasing with the annealing temperature. The interpretation of the $(\sqrt{31} \times \sqrt{31})R \pm 9°$ structure has been controversial for a long time. According to some authors, the observed cell is a coincidence cell between an aluminium-rich cubic phase and the underlying hexagonal lattice (Chang, 1968; French and Somorjai, 1970). According to other authors, it corresponds to a compact piling of pairs of Al (111) planes separated by an hexagonal symmetry defect (Gautier *et al.*, 1994). This interpretation is reinforced by the observation of the same LEED pattern after an evaporation of two aluminium monolayers on the non-reconstructed (1 × 1) surface (Vermeersch *et al.*, 1990). It is also confirmed by a recent experiment using grazing incidence x-ray diffraction, which does not suffer the limitations of electron-based techniques, in particular multiple scattering effects (Renaud *et al.*, 1994). Structural studies have also been performed on the α-alumina ($\overline{1}$012) face. It displays (4 × 1), (3 × 1), (2 × 1) reconstruction patterns (Chang, 1971; Gillet *et al.*, 1987), associated with oxygen vacancy ordering along one of the surface directions. To the author's knowledge, no reconstruction has been observed on MgO(100) and there exists no quantitative modelling allowing an understanding of non-stoichiometric reconstruction mechanisms on insulating oxides.

2.6 Conclusion

Due to the variety and the complexity of oxide structures, the surfaces present many interesting features associated either with their chemical composition – various terminations for a given surface orientation, oxygen sub-stoichiometry – , or with their electrostatic properties – polar surfaces. We have put much emphasis on the microscopic parameters responsible for the surface relaxation and rumpling and on analytical models which reproduce their qualitative characteristics. However, the relative efficiency of the different mechanisms remains to be settled and checked by systematic experimental investigations. For example, the relaxation and rumpling in many oxide surfaces have not been determined experimentally, although they are predicted to be very strong in some cases. The same is true for local distortions around localized defects such as steps, kinks or oxygen vacancies. Many questions remain unsolved on the subject of non-stoichiometric reconstructions, whose unit cells have been determined, but the actual atomic positions are rarely known and no theoretical attempt to describe the mechanisms responsible for the vacancy ordering has been performed.

3

Electronic structure of surfaces

In the surface layers, the breaking of anion–cation bonds and the modifications of inter-atomic distances, induced by relaxation, rumpling or reconstruction effects, perturb the electrostatic potentials and the orbital hybridization. The surface electronic structure is thus modified and presents specific features compared to the bulk characteristics. In this chapter, we will discuss various aspects of these changes, relevant for planar semi-infinite systems, thin films and defected surfaces. We will restrict ourselves to the results obtained on single crystals, prepared under controlled conditions and studied in ultra-high vacuum (Henrich, 1983; 1985; Henrich and Cox, 1994). Instead of focusing on specific properties of this or that oxide, we will try to extract the general trends concerning the density of states, the gap width, the charge densities, etc., and, whenever this is possible, we will point out the physical origin of the differences found between the bulk and the surface electronic structure.

3.1 Experimental and theoretical studies

The study of the surface electronic structure requires specific tools. For example, in spectroscopic experiments, in order to enhance the surface signal with respect to the bulk one, to obtain information on the outer layers, one has to send or detect particles with small mean free paths, which mainly sample the few outer layers. Similarly, special care has to be taken in the numerical approaches. We will quickly review some aspects of this question, both from the experimental and the theoretical sides.

Experimental studies

Ultra-violet photoemission (UPS) gives information on the shape of the density of states in the valence band and on the presence of occupied surface states, but cannot yield the surface gap width nor give information on empty states. Many UPS experiments have been performed on insulating oxides (Henrich, 1985; Henrich and Cox, 1994). In most cases, within the experimental resolution, the shape of the surface valence band differs suprisingly little from the bulk one and intrinsic surface states have only been identified in the gap of a limited number of oxides. The weight of cation states in the valence band may be estimated by using resonant photoemission. In $TiO_2(110)$ (Zhang *et al.*, 1991), $SrTiO_3(100)$ (Brookes *et al.*, 1986) and $BaTiO_3(100)$ (Hudson *et al.*, 1993a), for example, it has proved that the valence band contains a non-negligible contribution of titanium states, which reveals a large part of covalent character of the oxygen–titanium bond.

X-ray photoemission provides the positions of the surface atom core levels if the photon energy is chosen in an appropriate energy range. This has been done for example on the SrO and TiO_2 faces of $SrTiO_3(001)$ (Brookes *et al.*, 1987; Courths *et al.*, 1990), and of $BaTiO_3(001)$ (Hudson *et al.*, 1993b). The analysis of the chemical shift into intra-atomic electron–electron repulsion and Madelung potential allowed to conclude that the surface ionic charges are lower and the surface anion–cation bond more covalent than in the bulk.

Electron energy loss spectroscopy characterizes the inter-band transitions between occupied and empty states and in particular the gap width. Surface specific transitions have been observed on MgO{100} (Henrich *et al.*, 1980; Didier and Jupille, 1994a) and confirmed by electron reflectance (He and Møller, 1986), on CaO{100} and on SrO{100} (Protheroe *et al.*, 1982; 1983) with energies smaller than the bulk gap width. In α-alumina, similar experiments display states in the gap, on both the (0001) and the ($10\bar{1}2$) faces (Gautier *et al.*, 1991b; Gillet and Ealet, 1992). On the contrary, in TiO_2 (Henrich *et al.*, 1976, Chung *et al.*, 1977; Tait and Kasowski, 1979), $BaTiO_3$ (Hudson *et al.*, 1993b) and $SrTiO_3$ (Hirata *et al.*, 1994b), the gap is found to be free of surface states. Aside from the examples quoted above, most observations of gap states could actually be correlated to a sub-stoichiometry in the surface layers. We will come back to this point in the last section of this chapter.

In the case of narrow-gap oxides, scanning tunneling microscopy performed in the spectroscopic mode gives hints on the local nature of the top of the valence band or of the bottom of the conduction band. For example, a local gap width of about 6 eV was determined on the SrO face of $SrTiO_3$, a value closer to that of bulk SrO (6.5 eV) than to that of bulk $SrTiO_3$ (about 3 eV) (Liang and Bonnell, 1994). This is in qualitative agreement with photoemission results on the SrO- and TiO_2-terminated surfaces (Hikita *et al.*, 1993; Hirata *et al.*, 1994b).

Theoretical studies

The main numerical approaches allowing a calculation of the bulk electronic structure of oxides have been recalled in the first chapter. Treating surface effects requires some care, especially concerning the geometry of the systems

Most of the first calculations have used an embedded cluster geometry, with a very limited number of atoms, typically less than ten. It was realized that a correct modelling requires that the Madelung potential created by the ions located outside the cluster are taken into account. Otherwise, errors in the electronic structure and in the energies result, in particular for the atoms located on the cluster edges (Sugano and Shulman, 1963; Sauer, 1989; Pacchioni *et al.*, 1992b). The usually low number of atoms considered in embedded cluster techniques allows a use of advanced quantum chemistry methods, and in particular a better account of electron correlations (Tsukada *et al.*, 1983), but this is done to the prejudice of long-range processes, such as band formation and screening effects.

The modelling of a single crystal surface by a slab, made of several layers, allows a description of periodic phenomena in the surface plane, but the two faces of the slab have to be distant enough not to interact. This geometry is now more and more used, because it is not limited by size effects in the directions parallel to the surface. The surface may also be modelled as an extended defect which scatters the electrons. A Green's function formalism (Wolfram *et al.*, 1973) or a transfer matrix method (Ivanov and Pollmann, 1980; 1981a; 1981b) then yield the electronic structure. Nevertheless these methods generally do not allow electron–electron interactions to be treated in a self-consistent manner.

Densities of states, band widths and, more generally, all quantities related to the electron delocalization are not very sensitive to the way electron–electron interactions are treated. On the contrary, charge densities, band positions and gap widths require a precise estimation of the Madelung potentials. This is particularly important in the case of polar surfaces but also on non-polar surfaces, when one wishes to compare the surface to the bulk, or several surfaces of different orientations. This is the reason why, in the following, we will only discuss the results obtained by self-consistent methods.

The understanding of a surface electronic structure involves two steps. First, assuming a rigid truncation of the bulk lattice, various effects are induced by the anion–cation bond breaking, whose importance depends on the surface orientation. In a second step, it is necessary to also take into account the relaxation of the inter-atomic distances in the outer layers, which further modifies the electronic structure. This is the presentation that will be followed in this chapter. Theoretical and experimental results may only be compared after the second step has been completed. We will first discuss the electronic characteristics of planar non-polar surfaces, before focusing on the specifics of polar surfaces. Finally, the modifications of the electronic structure induced by the presence of oxygen vacancies will be presented.

3.2 Electronic structure of non-polar surfaces

The anion–cation bond-breaking produced by the formation of a surface is responsible for several phenomena. Some have an electrostatic origin. This is the case for polarization which is induced by surface electric fields which are much larger than those in the bulk. This is also the case for shifts of the renormalized atomic energies of surface atoms and shifts of surface bands, which are induced by reduction of the Madelung potential at the surface. On the other hand, as far as covalent effects are concerned, a modification of the anion–cation hybridization takes place, due to the lower coordination of the surface atoms. This has important consequences on the gap width and on the electron distribution.

Madelung potentials and gaps

To quantify the value of the electrostatic potential acting on the atoms of a rigidly cut three-dimensional lattice, one has to extend the definition of the Madelung constant. In an ionic representation of a binary compound, the electrostatic potential which is exerted on an ion of type i, located in the nth plane – indexed from the vacuum side – may be written in the following way:

$$V_{i,n} = \frac{\alpha_{i,n} Q}{R} , \qquad (3.2.1)$$

with R the shortest anion–cation inter-atomic distance and Q the charge of the ions of opposite type. The $\alpha_{i,n}$ are the Madelung constants. They depend upon the depth of the nth layer and, in the general case, upon the type i of the atom under consideration. On non-polar surfaces, they may be estimated by plane-by-plane summations if the planes are neutral, or by summations involving larger repeat units, provided that they bear a zero net charge and no dipole moment (Massida, 1978). This procedure cannot apply to polar surfaces, which will be the subject of Section 3.3.

The electrostatic potential created by neutral planes decreases exponentially at long distances (Slater, 1970). This property was already used in Chapter 2, in the discussion of the electrostatic mechanism of surface relaxation. As a consequence, the plane-by-plane summation converges quickly and only a few layers in the vicinity of the surface have Madelung constants which differ from the bulk.

For example, in the rocksalt structure, in which anions and cations occupy equivalent sites in the unit cell, the $\alpha_{i,n}$ do not depend on i: along the $\{100\}$ orientation, the ratios α_n/α_∞, which give the relative value between the Madelung constants in the nth plane and in the bulk, are equal to $\alpha_1/\alpha_\infty = 0.963$, $\alpha_2/\alpha_\infty = 1.001$ and $\alpha_n/\alpha_\infty = 1.000$ for n larger than 2. Along the $\{110\}$ orientation, the ratios α_n/α_∞ are equal to 0.881, 1.018, 0.999 and 1.000, respectively, when n increases from 1 to 4 (Parry, 1975; Levine and Mark, 1966).

In both cases, the Madelung constant on atoms in contact with vacuum ($n = 1$) is smaller than in the bulk. The value of α_1 is a function of the surface coordination Z_s. It decreases as Z_s gets smaller. For example on the rocksalt$\{100\}$, $\{110\}$ and $\{211\}$ faces, on which the atoms are respectively five-fold, four-fold and three-fold coordinated ($Z_s = 5$, 4 and 3) – six-fold in the bulk – the ratios α_1/α_∞ are equal to 0.96, 0.88 and 0.65, respectively.

On atoms located deeper and deeper into the crystal, the Madelung constants converge to the bulk value with small oscillations. As Z_s decreases, the amplitude of the oscillations increases, and the bulk value is reached more and more slowly. The oscillations of $\alpha_{i,n}$ are due to the lack of neighbouring atoms in successive coordination shells.

Usually, the atoms which are located in the first layer have less first neighbours than in the bulk. Since their first neighbours bear charges of opposite signs, the electrostatic attraction is reduced ($\alpha_1 < \alpha_\infty$). On the contrary, on dense surfaces, atoms located in the second layer usually have all their first neighbours but their second coordination shell is incomplete. Since the latter exerts a repulsive interaction, a decrease in the number of second neighbours leads to a value of α_2 larger than α_∞.

In Born's model, in which the ionic charges have the same constant integer values in the bulk and at the surface, the reduction of the Madelung constants at the surface directly induces a reduction of the Madelung potential. However, when the electronic structure is more carefully considered, changes in the electronic distribution in the outer layers may also modify the Madelung potential. We will see in the following that, on most dense oxide surfaces, these modifications are small but, on more open surfaces, precise estimations have to be performed.

As was shown in Chapter 1, the Madelung potential renormalizes the atomic energies and shifts the anion and cation levels towards lower and higher energies, respectively (Equations (1.4.1) and (1.4.2)). In the surface layer, the effective levels of the cations are thus lower than in the bulk, and the reverse is true for the surface anions. The actual levels also depend upon the intra-atomic Hartree or Hartree–Fock terms, which shift the atomic levels in the opposite direction (Ellialtioglu *et al.*, 1978), but, generally, more weakly. The relative values of these two terms are fixed by the optical dielectric constant ϵ_∞ of the oxide (Chapter 4).

The shift of the surface atomic levels is reflected in a modification of the first moment of the local densities of states, i.e. on the centre of gravity of the valence and conduction bands (Equation (1.4.36) in Chapter 1). Fig. 3.1 shows this effect for several MgO surfaces. As the coordination number Z_s of the surface atoms decreases, the weakening of the Madelung potential is responsible for a raising of the valence band and a lowering of the conduction band, which is stronger on more open surfaces.

In Chapter 1, Equation (1.4.42), we have stressed that the gap width has two contributions, an ionic one which is a function of the energy separation between the anion and cation effective atomic levels, and a covalent contribution related to electron delocalization processes:

$$\varDelta = \sqrt{(\epsilon_C - \epsilon_A)^2 + 4F_{min}} . \tag{3.2.2}$$

At the surface, the above discussion shows that the ionic contribution $\epsilon_C - \epsilon_A$ is smaller than in the bulk. This acts towards a narrowing of the gap, the effect being larger on more open surfaces.

On the other hand, the covalent contribution $2\sqrt{F_{min}}$, which is a function of the lowest eigenvalue of H_{ND}^2 (H_{ND} the non-diagonal part of the Hamiltonian on an atomic orbital basis set), also depends upon the surface orientation. However, its variations with the latter are not easily predictable because the wave vector at which the gap opens or closes may be different at the surface and in the bulk.

The difficulty is exemplified when considering a two-dimensional square lattice with anions and cations at alternating positions along the two directions of the plane. When one assumes that cations have a single s outer orbital and that the anions have two degenerate p_x and p_y orbitals, in the infinite two-dimensional lattice, $F_{min} = 0$ by symmetry. When the lattice is truncated along the (10) direction to form a 'slab', each surface atom loses one neighbour. It is no longer possible to construct a wave function ψ_C, such that all the coefficients of ψ_A on the anion p_x and p_y orbitals simultaneously vanish. F_{min} is thus non-zero and the calculation shows that it is equal to the square of the anion–cation resonance integral $\beta_{sp\sigma}$ times a geometric factor which is a function of the number of arrays in the slab. Similarly, along the (11) orientation, on which surface atoms lose two first neighbours, $F_{min} = C'\beta_{sp\sigma}^2$.

In this simple example, the covalent contribution to the gap width does not vary in a monotonic way with the number of broken bonds. For some oxide structures, F_{min} may vanish in the bulk and not at the surface. In other cases, the surface value of F_{min} may be smaller than in the

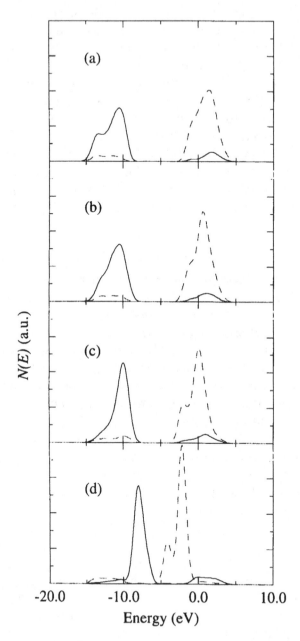

Fig. 3.1. Local density of states on anions (full line) and cations (dashed line) for MgO (arbitrary units). a) In the bulk; b) on the (100) face; c) on the (110) face; and d) on the (211) face (Goniakowski and Noguera, 1994a).

bulk. There is thus no systematic trend obeyed by F_{min} as a function of the surface orientation. Yet, in most cases, the variations of the ionic contribution to the gap seem to drive its changes.

It is striking that calculations of the surface electronic structure, using non-self-consistent methods, generally conclude that the surface gap width is of the same order as the bulk gap, while it is found to be reduced when self-consistent methods are used. This reduction is found, for example, in the calculations performed by Tsukada *et al.* (1983) and Gibson *et al.* (1992) on MgO(100), in those performed by Kasowski and Tait (1979) on TiO$_2$(110) and for a large range of surface orientations in various oxides (Goniakowski and Noguera, 1994a; 1994b). It should nevertheless be noted that, due to the systematic under-estimation of gap widths by *ab initio* LDA methods and their systematic over-estimation by *ab initio* Hartree–Fock methods, the value of the gap width is rarely discussed in recent advanced numerical studies. This is true in the bulk and *a fortiori* at surfaces.

One consequence of the surface gap narrowing is the existence of surface states connected to the top of the valence band and to the bottom of the conduction band, sketched in Fig. 3.2. The wave functions associated with these states have a propagating character in the surface layer but are exponentially damped inside the crystal.

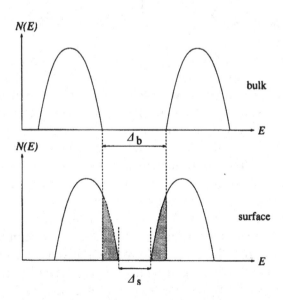

Fig. 3.2. Surface states at the top of the valence band and at the bottom of the conduction band and gap narrowing on a surface.

Surface density of states

The shape of the density of states reveals the peculiarities of the hybridization between anion and cation orbitals, which depend upon three parameters: the values of the resonance integrals; the coordination number of the surface atoms; and the energy separation between the relevant atomic levels. In the absence of relaxation, rumpling or reconstructions, the resonance integrals have the same values as in the bulk. The surface atom coordination numbers and the level separation, on the other hand, are smaller than in the bulk and they decrease as the surface becomes more open. We will first discuss how these modifications are reflected in the gross features of the local densities of states at the surface and, more specifically, in their second moments. Then we will focus on the details of the band shapes and on the possible occurrence of localized states in the gap.

Moments of the density of states   Within the assumption that electron delocalization takes place only between first neighbours, the second moment of the local density of states on an anion is related to its coordination number Z_A, to the effective anion–cation resonance integral β, to the anion atomic energies $\epsilon_{A\lambda}$ and to the degeneracy d_A of the outer levels (Equation (1.4.40) in Chapter 1):

$$\frac{M'_{2A}}{M_{0A}} = \frac{1}{d_A} \sum_{\lambda} \left(\epsilon_{A\lambda} - \frac{1}{d_A} \sum_{\lambda'} \epsilon_{A\lambda'} \right)^2 + Z_A \beta^2 . \qquad (3.2.3)$$

The first term is non-zero when the atomic levels which take part in the chemical bond are split in energy. It is generally larger at the surface than in the bulk, due to the lower symmetry of the crystal field. Yet, in most cases, it turns out to be much smaller than the second term, $Z_A \beta^2$, which is the covalent contribution to M'_{2A}. As a consequence of the lower coordination of surface atoms, the $Z_A \beta^2$ term yields a strong reduction of M'_{2A} at surfaces. The band-narrowing found in the local densities of states, e.g. those shown in Fig. 3.1, is thus of similar origin to the band-narrowing known on metal or semi-conductor surfaces. One should note that although, strictly speaking, M'_{2A} and M'_{2C} characterize the total width of the density of states (integration over the valence *and* conduction band energies), they also give the trend followed by the valence band width for M'_{2A} and by the conduction band width for M'_{2C}. This is clearly seen in Fig. 3.1. On dense surfaces, the band-narrowing is restricted to the surface layer. In the second atomic plane, the atoms usually have a complete first coordination shell, so that $Z_A \beta^2$ is equal to its bulk value.

Band shape The most important modifications in the shape of the local densities of states take place at the band edges. The states which result from the strongest anion–cation hybridization, found at the bottom of the valence band and at the top of the conduction band, are deeply modified by the bond-breaking at the surface. A typical case is presented in Fig. 3.3, for bulk and surface atoms on the (110) face of rutile TiO_2. Besides the relative shift of the surface and bulk bands discussed above, one observes a transfer of density weight from the bonding states to the non-bonding states at the top of the valence band. This effect is especially obvious on the bridging oxygens, which have the lowest coordination, in agreement with the calculations performed by Vogtenhuber *et al.* (1994). By contrast, the local density of states on surface cations, whether they are five- or six-fold coordinated, is close to the bulk one. A similar effect shows up in the modelling of $ZrO_2(001)$ (Orlando *et al.*, 1992). The changes in the local densities of states – band-narrowing and distortions – are not easily measurable by photoemission, which gives a signal integrated over all types of atoms and over a sample depth equal to the mean free path of the electrons. Scanning tunneling microscopy performed in the spectroscopic mode, on the other hand, can better reveal the details of the surface density of states in narrow-gap oxides.

Surface states in the gap Unlike compound semi-conductors, few oxides present surface states located deep in the gap and not connected to the valence or conduction band continuums. The reason probably lies in the fact that dangling bonds are not easily produced in compounds with octahedral environments. In tetrahedral semi-conductors, the bands result from the overlap of sp_3 hybrid orbitals, pointing along the bond directions. As a first approximation, on a given atom, each sp_3 orbital is hybridized with a single sp_3 orbital of a neighbouring atom. When the bond is broken, the orbital is left with a quasi-atomic character and is associated with a quantum state, usually located in the semi-conductor gap. In octahedral sites, on the other hand, each orbital is hybridized with several neighbouring orbitals. For example in MgO, the oxygen p_z orbital overlaps two magnesium orbitals located along the z-axis. If one bond is broken, the p_z orbital still takes part in delocalized states, because electron hopping can take place on the remaining bond. Well-defined surface states in oxide gaps nevertheless exist, whenever the surface coordination is so deeply reduced that a quasi-atomic state is formed. For example, on the (0001) face of α-Al_2O_3, terminated by a plane of three-fold coordinated aluminiums, several calculations (Causà and Pisani, 1989; Ohuchi and Koyama, 1991; Guo *et al.*, 1992; Godin and LaFemina, 1994a) have

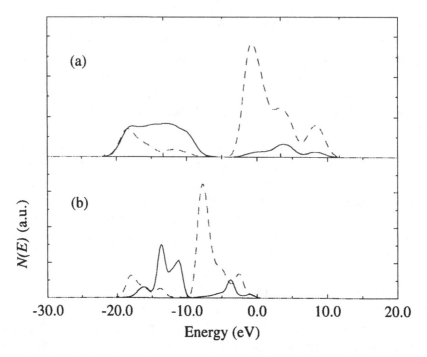

Fig. 3.3. Local densities of states: a) in the bulk; and b) on the (110) surface of TiO_2 (arbitrary units). Full and dashed lines are used for anion and cation densities of states, respectively (Goniakowski and Noguera, 1994a).

shown that there exists an empty surface state a few eV below the conduction band. A similar situation, although less well defined, exists on the (001) face of rutile (Kasowski and Tait, 1979). These intrinsic surface states should not be confused with the gap states induced by surface defects, such as steps, oxygen vacancies, etc., which will be considered later.

Surface charges

In mixed iono–covalent materials, the bond-breaking at the surface and the modifications of the Madelung potentials induce a charge redistribution. The latter cannot be directly determined experimentally. Some hints on the variations of the charge transfer in the surface planes may be obtained by an analysis of oxygen Auger lines or core level shifts. For example, on the TiO_2 face of $SrTiO_3(001)$, Courths *et al.* (1990) concluded from measurements of core level shifts that the surface ionic charges are lower and the surface anion–cation bond more covalent than in the bulk.

From a theoretical point of view, several authors find an increase of covalency at the surface. For example, Tsukada *et al.* (1983) obtain a reduction of the ionic charge on MgO(001) (1.54 versus 1.67 in the bulk), a reduction of 9% of the titanium charge on the (001) face of $SrTiO_3$, and also a strong reduction of the rhenium charge on ReO_3(001). They assign this systematic effect to an increase in the bond covalency between the surface atoms and the atoms in the underlying layers. On $SrTiO_3$(001), Ellialtioglu *et al.* (1978) also find a 0.45 electron charge reduction on the surface titanium ions, which they relate to a lowering of the cation effective atomic levels. Toussaint *et al.* (1987), on the other hand, using a non-self-consistent approach, obtain an increase of the titanium charge on this surface and an increase of the oxygen charge on the SrO-terminated face. At variance with the previous examples, some authors conclude that the surface charge density is very close to the bulk one. This is the case on MgO(100) by the *ab initio* Hartree–Fock method (Causà *et al.*, 1986), on α-Al_2O_3(0001) by the LDA method (Manassidis *et al.*, 1993), and for several other oxide surfaces (Goniakowski and Noguera, 1994a).

It is interesting to understand the mechanism of charge redistribution on clean planar surfaces because it also applies to other geometries in which the atoms have incomplete coordination shells – ad-atoms on surfaces; steps; or clusters. For this purpose, it is useful to write down a simple analytical expression for the ionic charges, as was done in Section 1.4 for the bulk charges. Yet, the tight-binding model used in the bulk relies on an assumption of orbital degeneracy for all anions (cations), which is not justified at the surface, because atoms of similar chemical types, located in different atomic planes, do not have the same effective atomic energies.

It is necessary to start from the more general formulation of a system, described by a Hamiltonian H, whose matrix elements on an atomic orbital basis set are of two types: the diagonal terms, equal to the effective atomic orbital energies ϵ_i and the non-diagonal terms which are non-zero when orbitals on two neighbouring sites are involved. H may thus be written as the sum of a diagonal operator H_D and a non-diagonal one H_{ND}. No hypothesis is made at this stage concerning the ϵ_i. Let G_0^+ and G^+ be the advanced Green's operators associated with H_D and H, respectively. G_0^+ is diagonal on the atomic orbital basis set, and its matrix elements are equal to:

$$\langle i|G_0^+|i\rangle = \frac{1}{E - \epsilon_i + i\eta} . \tag{3.2.4}$$

The density of states, proportional to the trace of the imaginary part of G_0^+ is a sum of delta peaks located at the energies ϵ_i of the atomic orbitals.

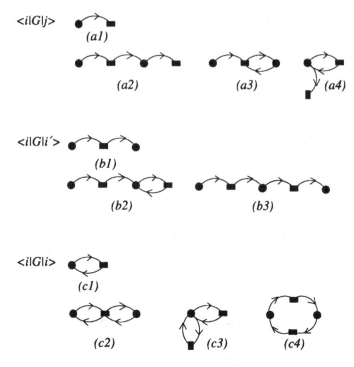

Fig. 3.4. Lowest-order diagrams associated with cation–anion orbital hybridiza-tion. Unlike the representations used in the many-body problem, the diagram vertices – a dot for an anion and a rectangle for a cation – here symbolize the matrix elements of G_0^+, while the arrowed lines indicate hopping processes associated with H_{ND}. Only 'bubble' diagrams of the *c1*- and *c3*-types are kept to derive the local charge transfer.

To estimate G^+, Dyson's equation is used:

$$G^+ = G_0^+ + G_0^+ H_{ND} G_0^+ + G_0^+ H_{ND} G_0^+ H_{ND} G_0^+ + ... \qquad (3.2.5)$$

The different terms involved in (3.2.5) may be represented by diagrams, the simplest of which are given in Fig. 3.4. To obtain a tractable expression for the charge, which generalizes Equation (1.4.48), only 'bubble' diagrams are kept but their summation to infinite order is performed. These diagrams represent hopping processes along closed paths, in which an electron, located on a given atomic orbital i, hops onto a neighbouring atom and then comes back to i. They provide a correct description of the *local* anion–cation hybridization inside a small cluster which includes the central atomic orbital i and all the first neighbours. Long-range band effects, on the other hand, are neglected, but the first and second moments of the local densities of states are exactly obtained.

Within this self-energy approximation, the Green's operator G^+ is diagonal on the atomic orbital basis set. Its matrix elements are equal to:

$$
\langle i|G^+|i \rangle = \frac{1}{E - \epsilon_i + i\eta}
$$

$$
+ \frac{1}{E - \epsilon_i + i\eta} \sum_j \frac{\beta_{ij}^2}{(E - \epsilon_i + i\eta)(E - \epsilon_j + i\eta)}
$$

$$
+ \frac{1}{E - \epsilon_i + i\eta} \left(\sum_j \frac{\beta_{ij}^2}{(E - \epsilon_i + i\eta)(E - \epsilon_j + i\eta)} \right)^2 + \dots
$$

$$(3.2.6)$$

i.e.:

$$
\langle i|G^+|i \rangle = \frac{1}{E - \epsilon_i + i\eta - \sum_j \frac{\beta_{ij}^2}{E - \epsilon_j + i\eta}} . \tag{3.2.7}
$$

It may be easily checked that (3.2.7) yields an electronic structure similar to that described in Chapter 1, when the j neighbours have the same atomic energy. For example, in this case, $\langle A|G^+|A \rangle$ reads:

$$
\langle A|G^+|A \rangle = \frac{1}{E - \epsilon_A + i\eta - \frac{Z_A \beta^2}{E - \epsilon_C + i\eta}} . \tag{3.2.8}
$$

The Green's matrix element has two poles, equal to:

$$
E = \frac{(\epsilon_C + \epsilon_A)}{2} \pm \frac{1}{2} \sqrt{(\epsilon_C - \epsilon_A)^2 + 4Z_A \beta^2} . \tag{3.2.9}
$$

The expressions of the local and total densities of states, derived in Equations (1.4.45) and (1.4.46), and of the ionic charges, derived in Equation (1.4.48), are then recovered. When the anion (cation) orbitals are degenerate, the self-energy approximation derived here is thus equivalent to writing $M(F)$ as a single delta function peaked at the position of its first moment.

However, (3.2.7) is more general and allows surface effects to be treated, i.e. a geometry in which the neighbours located in the surface plane or in the sub-surface layers have different atomic energies due to the different Madelung potentials that they suffer. The poles of (3.2.7) no longer have a simple expression, and the same is true for the ionic charges. However, a qualitative estimate is still feasible. For example, let us consider an atomic orbital on a surface anion of energy ϵ_{As}, hybridized with Z_1 cation orbitals of atomic energy ϵ_1 in the surface layer and Z_2 orbitals of atomic energy ϵ_2 in the underlying plane ($Z_1 + Z_2 = Z_{As}$, the surface anion coordination number). $\langle A|G^+|A\rangle$ is then equal to:

$$\langle A|G^+|A\rangle = \frac{1}{E - \epsilon_A + i\eta - \frac{Z_1\beta^2}{E-\epsilon_1+i\eta} - \frac{Z_2\beta^2}{E-\epsilon_2+i\eta}} \cdot \qquad (3.2.10)$$

Up to second order in the energy difference $(\epsilon_1 - \epsilon_2)$, the pole equation may be approximated by:

$$E - \epsilon_{As} - \frac{(Z_1 + Z_2)\beta^2}{E - \overline{\epsilon_{Cs}}} = 0, \qquad (3.2.11)$$

with an average cation energy $\overline{\epsilon_{Cs}}$ defined by:

$$\overline{\epsilon_{Cs}} = \frac{Z_1\epsilon_1 + Z_2\epsilon_2}{Z_1 + Z_2} \cdot \qquad (3.2.12)$$

The local charge transfer, associated with each kind of orbital, is thus given by an expression similar to Equation (1.4.48), as a function of the dimensionless parameter $Z_{As}\beta^2/(\overline{\epsilon_{Cs}} - \epsilon_{As})^2$. Ionicity increases as this ratio gets smaller. The variations of the charge transfer with respect to the bulk value are fixed by the relative values of $Z_{Ab}\beta^2/(\epsilon_{Cb} - \epsilon_{Ab})^2$ and $Z_{As}\beta^2/(\overline{\epsilon_{Cs}} - \epsilon_{As})^2$. These two quantities differ because the surface coordination number Z_{As} is smaller than the bulk one Z_{Ab}. This effect, taken alone, would induce an increase of ionicity. However, the mean energy difference $\overline{\epsilon_{Cs}} - \epsilon_{As}$ is also smaller at the surface than in the bulk, due to the reduction of the Madelung potential on surface atoms. This points towards a decrease in the surface charges. The latter effect was invoked by Tsukada *et al.* (1983) and Ellialtioglu *et al.* (1978). It is lacking in the non-self-consistent calculations. In this case, the first effect alone is present and the ionicity increases.

The competition between both effects – reduction of the number of bonds for electron delocalization and decrease of the atomic energy difference – leads to a rough balance when the electron numbers are summed over all orbitals. This is true on many non-polar oxide surfaces, even when a large number of bonds are broken (Goniakowski and Noguera, 1994a; 1994b). However, the two effects act differently on the atomic orbitals, depending upon their directionality with respect to the surface. This is exemplified by the MgO(001) surface, on which the atoms lose one neighbour in the z-direction: the reduction of Z prevails for the p_z orbital and the reduction of $\epsilon_C - \epsilon_A$ for the p_x and p_y orbitals. The electron number on the magnesium s level, on the other hand, is roughly unchanged, because the s orbital is more isotropic and a compensation between the five bonds takes place (Russo and Noguera, 1992a).

As a general statement, therefore, the actual charge redistribution depends upon several characteristics of the surface, such as the number of broken bonds, the strength of screening effects in the material, etc. In the calculations, the assumptions which are made are of crucial importance. In particular, screening processes have to be treated accurately, which requires a self-consistent account of the electron–electron interactions. In addition, a substantial number of atoms has to be included in the electronic structure calculation in order that delocalization effects are well described and in order that the screening cloud is not limited by the cluster size. All these remarks explain the controversies which remain in the literature concerning the ionicity of oxygen–cation bonds at the surface.

A complete description of charge redistributions is given by charge maps. They are generally obtained by *ab initio* methods, and one may compare the anion–cation electron sharing at the surface and in the bulk. For example, in α-alumina, the effect of back-donation of electrons from the oxygen anions to the aluminium cations was shown to be more pronounced at the surface than in the bulk (Causà and Pisani, 1989). The distortions of the atomic orbitals induced by the surface electric field may also be estimated. This polarization effect is not included in the tight-binding methods.

The above analysis may be extended to understand the local electronic structure on any kind of under-coordinated atom, for example on step edges or on kinks. Stepped or kinked surfaces involve atoms with many non-equivalent environments, which are characterized by the numbers Z_1 and Z_2 of atoms in their first and second coordination shells. As a general statement, the Madelung potential decreases strongly as Z_1 decreases, as already discussed for planar surfaces, and decreases more slowly as Z_2 increases. These variations are legitimate, since first and second neighbours have charges of opposite sign. The effective cation atomic levels are pushed towards lower energies, and the anion levels towards higher energies as the Madelung potential gets smaller. This may result in the existence of gap states. Thibado *et al.* (1994), for example, have observed such states at the edges of ZnO terraces, using an STM. Similarly a transition at an energy lower than the gap, found by UV diffuse reflectance in MgO, was assigned to the presence of steps and kinks (Garrone *et al.*, 1980). Thus, as in the case of planar surfaces, the local electronic gap becomes smaller and smaller as Z_1 decreases. The shift of the atomic levels suggests that steps or kinks may have a higher reactivity than terrace sites: a larger basicity for the oxygens and a larger acidity for the surface cations. This is exemplified by the theoretical study of water dissociation on terraces or steps of MgO, which concludes that the adsorption is dissociative on step edges but not on the terraces (Langel and Parrinello, 1994). The higher activity of stepped or vicinal surfaces has been noted by several authors, among which are Onishi *et al.* (1987) and Brookes *et al.* (1987) on MgO and $SrTiO_3$, respectively.

On the other hand, Muryn *et al.* (1991) have observed that the presence of steps and point defects on $TiO_2(100)$ does not change the ability of the surface to dissociate water molecules. The relationship between the local electronic structure and the reactivity will be made explicit in Chapter 6.

As far as charges on under-coordinated atoms are concerned, the same competition between covalent effects (reduction of $Z\beta^2$) and electrostatic effects (reduction of $\epsilon_C - \epsilon_A$) is at work. Studies of planar or defected surfaces with atoms of similar coordination number yield similar surface charges (Goniakowski and Noguera, 1995c). Precise charge maps around under-coordinated atoms on MgO(100) surface have been obtained recently by a density functional approach taking into account atomic relaxation (Kantorovich *et al.*, 1995).

To conclude, charge redistributions around under-coordinated atoms are subtle effects. They generally include two mechanisms: a reduction of electron delocalization due to the surface bond-breaking, and a reduction of the local mean anion–cation energy difference. The first effect increases the ionicity while the second decreases it. Even when both effects balance each other, i.e. when the surface charge is close to that of the bulk, the electron numbers on the outer orbitals of the surface atoms are modified. Polar or weakly polar surfaces do not follow these trends. They will be considered in the next section.

Electronic characteristics of thin films

In very thin films, new effects may take place, because the finite thickness is responsible for further modifications of the Madelung potential. This shows up, for example, when one considers unsupported MgO films, with thicknesses n ranging from 1 to 6 atomic planes and several orientations ((100), (110) and (211)). As a first gross approximation, the atoms may be assumed to remain at their bulk positions. The application of the self-consistent tight-binding method yields the gap width Δ and the ionic charges Q_s borne by the surface atoms, as a function of their coordination number Z_s. The results are as follows:

- (001) orientation

$$n = 1 \qquad Z_s = 4 \qquad \Delta = 5.5 \text{ eV} \qquad Q_s = 1.25$$
$$n = 2 \qquad Z_s = 5 \qquad \Delta = 6.8 \text{ eV} \qquad Q_s = 1.24$$
$$n = 3 \qquad Z_s = 5 \qquad \Delta = 6.7 \text{ eV} \qquad Q_s = 1.25$$

- (110) orientation

$$n = 1 \qquad Z_s = 2 \qquad \Delta = 4.3 \text{ eV} \qquad Q_s = 1.10$$
$$n = 2 \qquad Z_s = 4 \qquad \Delta = 6.7 \text{ eV} \qquad Q_s = 1.24$$
$$n = 3 \qquad Z_s = 4 \qquad \Delta = 5.8 \text{ eV} \qquad Q_s = 1.23$$
$$n = 4 \qquad Z_s = 4 \qquad \Delta = 6.3 \text{ eV} \qquad Q_s = 1.23$$

- (211) orientation

$$n = 2 \qquad Z_s = 2 \qquad \Delta = 2.4 \text{ eV} \qquad Q_s = 0.34$$
$$n = 3 \qquad Z_s = 3 \qquad \Delta = 4.0 \text{ eV} \qquad Q_s = 1.25$$
$$n = 4 \qquad Z_s = 3 \qquad \Delta = 0.3 \text{ eV} \qquad Q_s = 1.22$$
$$n = 5 \qquad Z_s = 3 \qquad \Delta = 1.5 \text{ eV} \qquad Q_s = 1.25$$
$$n = 6 \qquad Z_s = 3 \qquad \Delta = 1.9 \text{ eV} \qquad Q_s = 1.25$$

The gap width Δ presents large variations as a function of n, especially along the (211) orientation. However, when the film thickness exceeds one repeat unit in the direction perpendicular to the surface, the electronic structure on surface atoms tends towards that of a semi-infinite system. The variations of Δ are driven by the values of the Madelung potential. Along (211), this latter varies non-monotonically with n, in a way which may be understood by counting how many atoms belong to the first and second coordination shells of the surface atoms.

The surface charges Q_s turn out to be very close to the bulk charges. The rough balance between electrostatic and covalent effects invoked for semi-infinite planar surfaces takes place in most cases, except for the smallest thicknesses in the less-dense orientations, when the coordination number is smaller than or equal to 2.

The values of Δ are fixed by the positions of the renormalized atomic levels of the surface atoms, which partly determine the transport properties of these films, but also their chemical reactivity. Fig. 3.5 shows the large variations of the lowest cation ϵ_C and highest anion ϵ_A renormalized atomic energies, as a function of the film thickness n, for different orientations.

These considerations suggest that very thin oxide films may present specific electronic properties, which might find interesting applications, for example in the field of micro-electronics or opto-electronics, and an enhanced reactivity similar to that of oxide powders. This point is more and more recognized in the literature. Thin oxide films may be grown on metallic substrates by deposition of metallic atoms and subsequent oxidation. By this method, many crystalline films have been obtained, for example BaO on W(110) (Shih *et al.*, 1988; Mueller *et al.*, 1990) – a

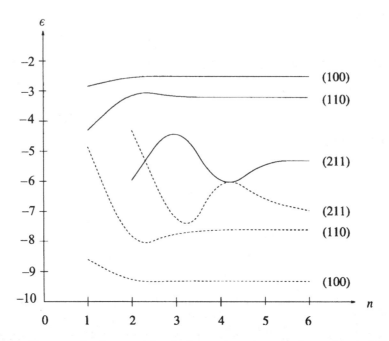

Fig. 3.5.   Renormalized energy levels for cations (magnesium s level: full line) and for anions (the highest oxygen p level: dashed line) at the surface of thin unsupported MgO films, as a function of their number of layers n, for three surface orientations. Continuous lines are drawn to guide the eye.

system which has been modelled by Hemstreet *et al.* (1988) – ZrO_2 on Pt(111) (Maurice *et al.*, 1990), MgO on Mo(100) (Wu *et al.*, 1991b), FeO on Pt(111) (Galloway *et al.*, 1993), Fe_3O_4 on Pt(111) (Weiss *et al.*, 1993; Barbieri *et al.*, 1994), Na_2O on Pd(100) (Onishi *et al.*, 1994), Al_2O_3 on Ta(110) (Chen and Goodman, 1994), NiO on Au(111) (Hannemann *et al.*, 1994), TiO_2 or Ti_4O_7 on Pt(111) (Boffa *et al.*, 1995). Interesting questions such as the decrease of the metal work-function upon deposition of an oxide (Shih *et al.*, 1988), the modification of the critical temperatures of phase transitions in confined geometries (Frederick *et al.*, 1991b; Gassman *et al.*, 1994) and the comparison of the reactivity with that of powders (Frederick *et al.*, 1992) have been raised. Electronic states at the interface, growth modes and conditions of epitaxy, on the other hand, rely on parameters which are very similar to those relevant in the inverse system: metal on oxide, which will be the subject of Chapter 5.

Surface energies

The surface energy is the energy required to create a surface. It is a positive quantity. In insulating oxides, its order of magnitude ranges from

0.5 to 1 J/m^2, for the most stable surfaces. It is very difficult to measure and few experimental results are available (Jura and Garland, 1952; Livey and Murray, 1956).

Until recently, most surface energy calculations have been based on classical atomistic approaches, which emphasize the competition between Madelung interactions and short-range repulsion between pairs of atoms. The surface energy may then be related to the value of the cohesion energy in the bulk and to the surface geometry. In Born's model, in the bulk, the total energy per formula unit reads:

$$E = -\frac{\alpha_b Q^2}{R} + \frac{Z_b A}{R^n} , \qquad (3.2.13)$$

as a function of the Madelung constant α_b and of the coordination number Z_b. The cohesion energy is obtained by minimizing E with respect to R:

$$E_{coh} = \frac{\alpha_b Q^2}{R_0} \left(1 - \frac{1}{n}\right) . \qquad (3.2.14)$$

In the absence of structural distortions, the surface energy E_s is determined by the values of the Madelung constant α_s and of the coordination number Z_s on surface atoms. It is equal to:

$$E_s = E_{coh} \left(1 - \frac{\alpha_s/\alpha_b - Z_s/nZ_b}{1 - 1/n}\right) . \qquad (3.2.15)$$

It is an increasing function of the number of broken bonds $Z_b - Z_s$. As in the case of metal surfaces, dense surfaces are the most stable. The surface energies of rocksalt oxides (Mackrodt, 1988; 1989), UO_2 and corrundum oxides (Mackrodt et al., 1987) have been calculated by atomistic methods, primarily based on Born's model, but including also polarizability corrections in the shell model. The surface free energy has also been modelled for three SnO_2 surfaces: (001), (211) and (110), taking into account the phonon degrees of freedom in the Einstein model (Mulheran and Harding, 1993).

In quantum approaches the complexity increases since both covalent and electrostatic effects contribute to the surface energy. The expression for the covalent energy presented in Chapter 1, Equation (1.4.62):

$$E_{cov} = \frac{-4n_0 Z_A \beta^2}{\sqrt{(\epsilon_C - \epsilon_A)^2 + 4Z_A \beta^2}} , \qquad (3.2.16)$$

may be generalized to estimate the covalent contribution to the surface energy, in a way similar to what has been done for the charges in Section 3.2, Equations (3.2.10) to (3.2.12). Despite a smaller value of the energy difference $\epsilon_C - \epsilon_A$ at the surface, the variations of E_{cov} are driven by the value of the surface coordination number Z_{As}. The covalent

contribution to the surface energy is thus positive, and it increases with
the number of broken bonds. The relative weight of electrostatic and
covalent effects on a surface is unknown and will likely be the subject of
further work in the future.

Calculations of surface energies by quantum methods have been performed for a number of surfaces, such as the MgO (100), (110), (211)
or (111) surfaces (Causà *et al.*, 1986; Gibson *et al.*, 1992; Goniakowski
and Noguera, 1994a), the (0001) and ($\bar{1}102$) surfaces of α-Al$_2$O$_3$ (Guo *et
al.*, 1992; Manassidis *et al.*, 1993), and various terminations of SnO$_2$(111)
(Godin and LaFemina, 1994b).

Consequences of structural distortions on the electronic structure

Up to this point, we have investigated the consequences of bond-breaking
on the surface electronic properties, assuming that the atoms remain
located at bulk-like positions. However, the structural distortions which
take place in the outer layers modify the inter-atomic distances and require
a new consideration of all the previous arguments.

Madelung potential   As far as inter-atomic distances are concerned, the
inward relaxation discussed in the preceding chapter induces a contraction of several anion–cation inter-atomic distances around each surface
atom and thus a reinforcement of the surface Madelung potential. In
the rocksalt structure, relaxation effects do not invert the hierarchy of
Madelung potentials for the different surface orientations (Goniakowski
and Noguera, 1995a). The same is true on the (111), (110) and (310)
faces of CeO$_2$ (Sayle *et al.*, 1994) and on SnO$_2$ (Mulheran and Harding,
1993). For more complex structures, such as corundum oxides, deeper
modifications of the Madelung potentials take place (Mackrodt *et al.*,
1987; Blonski and Garofalini, 1993).

Gap width and surface states   The above-mentioned reinforcement of
the Madelung potential pushes the surface states and the surface band
edges closer to the bulk band edges and re-opens the gap. Depending
upon the oxide, the surface gap width remains smaller than in the bulk
or is indistinguishable from the latter (Goniakowski and Noguera, 1995a).
Electron energy loss experiments have evidenced electronic excitations at
energies smaller than the bulk gap width in MgO{100} (Henrich *et al.*,
1980; Didier and Jupille, 1994a), CaO{100} and SrO{100} (Protheroe *et
al.*, 1982; 1983). This proves that the rocksalt{100} surfaces belong to the
first family. At the surface of vanadium pentoxide, on the other hand,
the gap width is equal to that of the bulk (Zhang and Henrich, 1994), as
is also the case for rutile(110). Kasowski and Tait (1979) were the first

authors to assign this result to a balance between the gap reduction on unrelaxed surfaces and the effect of surface relaxation. They estimated that an inward relaxation of the bridging oxygens by 0.2 Å, leading to a Ti–O distance contraction of 0.1 Å, would destroy the gap states and bring the gap width back to its bulk value. However, they did not perform an actual optimization of the geometry. Several recent works have done the full optimization and checked the validity of the idea (Vogtenhuber *et al.*, 1994; Ramamoorthy *et al.*, 1994; Goniakowski and Noguera, 1995a). The position of the surface states which are not connected to the band edges is also modified, with a tendency to merge into the bulk bands. In α-Al$_2$O$_3$(0001), a controversy exists concerning the energy of the surface state in the relaxed geometry (Godin and LaFemina, 1994a; Manassidis *et al.*, 1993; Causà and Pisani, 1989).

Surface charges   Inward relaxation effects have two antagonistic effects on the ionic charges. They increase the resonance integrals and increase the energy difference between anion and cation effective levels. Since the charges depend upon the ratio $Z_{As}\beta^2/(\epsilon_{Cs} - \epsilon_{As})^2$, in most cases no noticeable variation results.

Consequences of the relaxation on the relative stability of different surfaces and on the crystal morphology   Relaxation effects always stabilize the surface and induce a lowering of the surface energy. This lowering involves a covalent part – associated with the increase in the resonance integrals – and an electrostatic part – associated with the increase in the Madelung interactions. On dense surfaces, which are weakly relaxed, the energy lowering is small. However, there exist examples in the literature where it reaches 30–50% of the total energy. When such cases occur, relaxation effects have to be carefully taken into account to predict the relative stability of different faces of a given material. This point is well exemplified by the work of Mackrodt *et al.* (1987) and Lawrence *et al.* (1988) on corundum oxides: α-Al$_2$O$_3$, α-Fe$_2$O$_3$ and α-Cr$_2$O$_3$, using an atomistic approach which includes polarization corrections in the shell model and non-coulombic pair interactions between first, second and third neighbours. They find that relaxation effects lower the surface energy by about a factor of two, in agreement with a recent *ab initio* quantum calculation on α-Al$_2$O$_3$ (Manassidis *et al.*, 1993). In the absence of relaxation, increasing surface energies are found in the order:

$$(10\bar{1}2) < (11\bar{2}0) < (10\bar{1}1) < (0001) < (10\bar{1}0) .$$

When relaxation effects are taken into account, the order is changed, to become, for α-Al$_2$O$_3$:

$$(0001) < (10\bar{1}0) \approx (10\bar{1}2) < (11\bar{2}0) < (10\bar{1}1)$$

and, for $\alpha\text{-Fe}_2\text{O}_3$:

$$(10\bar{1}2) \approx (0001) < (11\bar{2}0) < (10\bar{1}0) \approx (10\bar{1}1) \,.$$

This classification is in agreement with the known stability of the corundum (0001) and $(10\bar{1}2)$ faces.

The stability of crystal surfaces determines the morphology of crystal growth (Hartman and Perdok, 1955). The growth rate of an (h, k, l) face increases in proportion to the energy which is released when a monolayer is added to the growing face; this energy is itself proportional to the surface energy (Hartman and Bennema, 1980). Numerical results have nevertheless to be compared with experimental results with much caution, because they implicitly assume ultra-high vacuum conditions which are not fulfilled during the crystal growth. Mackrodt *et al.* (1987) have calculated growth shapes expected for $\alpha\text{-Al}_2\text{O}_3$ and $\alpha\text{-Fe}_2\text{O}_3$, on the basis of relaxed and unrelaxed surface energies. Relaxation effects are responsible for the differences found between $\alpha\text{-Al}_2\text{O}_3$ and $\alpha\text{-Fe}_2\text{O}_3$ and, more specifically, for the prevalence of $\left\{10\bar{1}0\right\}$ basal planes in the first-named oxide (Fig. 3.6).

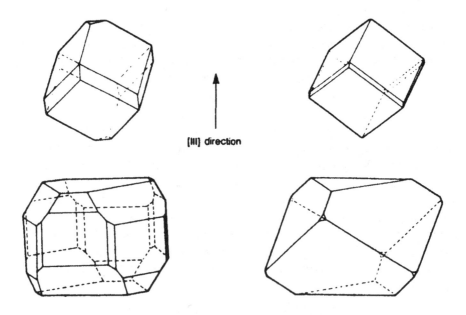

Fig. 3.6.   Growth shapes of $\alpha\text{-Al}_2\text{O}_3$ (*left*) and $\alpha\text{-Fe}_2\text{O}_3$ (*right*), according to Mackrodt *et al.* (1987). The shapes in the upper part of the figure are calculated assuming unrelaxed geometries. Relaxation effects were taken into account to derive the shapes in the lower part of the figure.

3.3 Polar surfaces

Polar surfaces are type 3 surfaces, according to Tasker's (1979a) classification. They present distinctive features because of the presence of a net charge and a net dipole moment in the repeat unit, in a direction perpendicular to the surface.

Polar and weakly polar surfaces

Polar surfaces generally have a very low stability and, for this reason, they have not been studied as thoroughly as non-polar surfaces.

In rocksalt oxides, the polar face with the lowest indices is the $\{111\}$ face. It presents sequences of alternating and equally spaced anion and cation planes. The planar face has a high surface energy. In nearly all experimental works done until now, it turned out to be either faceted or non-stoichiometric. The planar surface is more easily obtained when a thin oxide film is grown on a substrate.

- Floquet and Dufour (1983) studied the $NiO\{111\}$ surface. It was $(\sqrt{3} \times \sqrt{3})R30°$ reconstructed, and was likely to be stabilized by silicon impurities. Warren and Thiel (1994) grew it as a thin film on a nickel substrate. Rohr *et al.* (1994) found that the p(2×2) reconstruction on the bare NiO(111) surface is destoyed under hydroxylation of the surface. The p(2×2) reconstruction was also observed by Ventrice *et al.* (1994) and described as due to the formation of micro-pyramids, a geometry predicted by Wolf (1992) to lead to a larger stabilization.

- In Henrich's (1976) and Onishi *et al.*'s (1987) experiments, the polar $MgO\{111\}$ face was faceted. Henry and Poppa (1990) could prepare planar $MgO\{111\}$ thin films of a few layers thickness. Gajdardziska-Josifovska *et al.* (1991), on the other hand, observed a $(\sqrt{3} \times \sqrt{3})R30°$ reconstruction on a semi-infinite, oxygen-annealed MgO sample. They discarded the possibility of faceting or impurity contamination. The reason for the surface stabilization in this experiment remains unclear.

- Hydroxylated polar CoO(111) surfaces have also been prepared and studied (Hassel and Freund, 1995).

In the wurtzite structure, the polar faces are the (0001) and (000$\bar{1}$) faces. In ZnO, they are terminated by a zinc and an oxygen plane, respectively. It seems that their surface energies are lower than in the rocksalt crystals. The experiments performed on these materials do not mention impurity or faceting problems, and the Zn-terminated surface was found to be unrelaxed (Sambi *et al.*, 1994).

The $Cu_2O(100)$ polar surface has also been studied (Schulz and Cox, 1991). It is $(3\sqrt{2} \times \sqrt{2})R45°$ reconstructed when it is terminated by copper atoms and unreconstructed otherwise.

The (111) polar face of Fe_3O_4 presents a deficiency of iron atoms in its outer layer (Barbieri *et al.*, 1994).

One should note that some surfaces, generally regarded as type 1 or type 2 surfaces, are actually polar surfaces. Tasker's classification relies on an ionic description of the materials. But, as we have seen, the ionic charges are not equal to the formal charges. As a consequence, the actual net charge and dipole moment in the repeat unit may be quite different from that estimated with the formal charges. Let us, for example, consider the {001} face of a ternary oxide such as $SrTiO_3$. In the bulk, the neutrality of the unit cell requires that $Q_{Sr} + Q_{Ti} + 3Q_O = 0$, but the titanium and strontium charges do not necessarily fulfill the relationship $Q_{Ti} = 2Q_{Sr}$, at variance with the ionic models. If one notes $Q = Q_{Sr} + Q_O = -Q_{Ti} - 2Q_O$, the {001} surface, which presents sequences of SrO and TiO_2 planes, turns out not to be of type 1, as predicted by Tasker's classification, but of type 3, each plane parallel to the surface bearing a charge equal to $\pm Q$ per unit cell, which yields a non-zero dipole moment in the repeat unit. We will call such surfaces, *weakly* polar surfaces.

Electrostatic arguments

A semi-infinite crystal cut along a polar direction has an infinite electrostatic energy, because the electrostatic field has a non-zero mean value in the material. Yet, under some specific conditions that we will now consider, the macroscopic field may be cancelled out and the surface stabilized. The following arguments rely upon a macroscopic analysis, which means that:

- The discrete atomic structure inside each layer parallel to the surface is neglected. One takes into account only the uniform component (term $G_\parallel = 0$ in the Fourier series) of the electrostatic potential.
- The electron density is assumed to be localized in the planes parallel to the surface, with no inter-plane overlap. The arguments thus strictly apply in the limit in which the inter-plane spacing is large compared to the atomic sizes.

Macroscopic electric field   Let us consider a slab of equi-distant layers, bearing positive and negative charge densities equal to $\pm \sigma$ (Fig. 3.7a). The repeat unit has a dipole moment density equal to σR. As for an association of capacitors, Gauss' theorem allows an estimation of the electric field $\mathscr{E}$ perpendicular to the layers and of the electrostatic potential V. The field $\mathscr{E}$ turns out to be equal to zero between two double-layers and equal to $4\pi\sigma$ inside a double-layer. Its mean value in the slab is, therefore, non-zero:

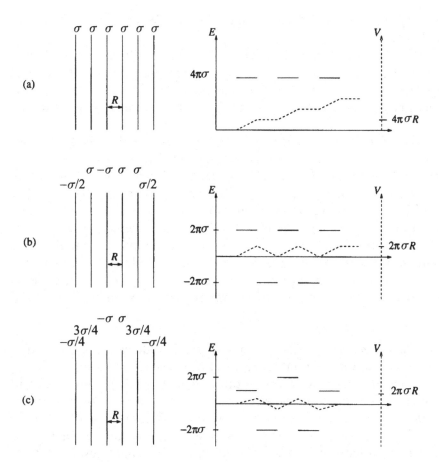

Fig. 3.7. Spatial variations of the electric field $\mathscr{E}$ (full lines) and of the electrostatic potential V (dashed line) in a thin sample cut along a polar direction. (a) When the planes bear charge densities equal to $\pm \sigma$, the electrostatic potential increases monotonically through the sample. (b) The charge density on the outer planes is corrected by $\sigma' = \sigma/2$: the electrostatic potential no longer increases proportionally to the sample thickness, but oscillates around a non-zero value. (c) A dipole moment $\sigma''R$ is added on the outer faces: the electrostatic potential oscillates around a zero value if $\sigma'' = \sigma/4$.

$\overline{\mathscr{E}} = 2\pi\sigma$. The electrostatic potential, on the other hand, increases from the left to the right of the slab in Fig. 3.7a, by an amount $\delta V = 4\pi\sigma R$ per double layer. δV is large, typically of the order of several tens eV. Unlike non-polar surfaces, it is no longer possible to use the concept of a surface Madelung constant, because the electrostatic potential is different from that of the bulk in *every* layer. The total electrostatic energy is proportional to the slab thickness. It is infinite for macroscopic systems and very large even for thin films. This is the origin of the surface instability.

In the following, we will consider two ways of cancelling the macroscopic field on the basis of purely electrostatic arguments. We will temporarily forget that the system has electronic degrees of freedom allowing spontaneous charge redistributions in response to the electrostatic potentials.

Surface charge modifications   Let us imagine that the charge densities on the two outer planes of the slab are not equal to $\pm\,\sigma$ but to $\pm\,(\sigma - \sigma')$ (Fig. 3.7b). According to Gauss' theorem, the electric field is zero outside the material, equal to $4\pi(\sigma - \sigma')$ inside each double-layer and equal to $-4\pi\sigma'$ in the space region between two double-layers. Its mean value is equal to $\overline{\mathscr{E}} = 2\pi(\sigma - 2\sigma')$. A reduction of the surface charge by a factor of two with respect to the bulk: $\sigma' = \sigma/2$, thus allows a cancellation of $\overline{\mathscr{E}}$ and suppresses the monotonic increase of the electrostatic potential. In this case, this latter fluctuates between 0 and $+ 2\pi\sigma R$. Yet, since the mean value of V is non-zero, the energy remains proportional to the slab thickness.

Addition of a surface dipole   An additional modification of the surface charges may lower the total energy. Let us consider that a dipole density $\sigma''R$ is added on the two faces of the system (Fig. 3.7c). Compared with the preceding estimation, the electrostatic field is changed only inside the double-layer on which the dipole has been placed. A value of $\sigma'' = \sigma'/2 = \sigma/4$ yields oscillations of V around zero, so that the macroscopic electrostatic energy of the system vanishes. A similar result could be obtained by a modification of the outer double-layer spacing δR, such that the dipole moment $\sigma(R - \delta R)/2 = \sigma''R$. This corresponds to an inward relaxation equal to $\delta R = R/2$.

The conditions on σ' and σ'' that have been just obtained may be generalized for slabs in which the inter-plane spacings are alternatively equal to R (in a double-layer) and R' (between two double-layers). They read: $\sigma' = \sigma R/(R + R')$ and $\sigma'' = \sigma R'/2(R + R')$. On the rocksalt$\{111\}$ face, SrTiO$_3\{100\}$ face and blende$\{100\}$ face, R' is equal to R. It is equal to $3R$ on the blende $\{111\}$ surface.

Discussion   The stabilizing charges may be induced by charged defects, such as oxygen vacancies, impurities (Floquet and Dufour, 1983; Cox and William, 1985) or pairs of charged defects. In the case of rocksalt oxides, for example, the saturation of surface sites by adsorbed protons and hydroxyl groups yields just the required charge $\sigma' = \sigma/2$.

Nosker *et al.* (1970) have proposed models of polar surfaces for the wurtzite$\{0001\}$ face and the blende$\{111\}$ face, in which a macroscopic number of surface ions spontaneously desorb to yield the appropriate surface charge density, and in which the surface vacancies order to form reconstructed (2 × 2) or (3 × 3) surface cells. The existence of ordered vacancies, inducing a surface reconstruction, was also proposed to account for the stabilization of blende semi-conductor compound polar faces (Tong *et al.*, 1984; Lannoo, 1993).

The stability criterion may be fulfilled by atomic redistributions involving several atomic planes: Nosker *et al.* (1970), for example, have imagined faceted structures for the wurtzite and blende polar faces. In the case of the rocksalt{111} face, Fig. 3.8 exemplifies two possible faceted structures associated with non-stoichiometric reconstructions. In Fig. 3.8a, every other atom is missing in the surface layer, so that parallel streaks are formed with (100) and (110) facets. The value of the electrostatic potential on the different layers is then given by Fig. 3.7b. In the second case (Fig. 3.8b), three out of every four atoms are missing in the surface layer and one out of four is missing in the second layer. Square facets of (100) orientations involving four atoms are formed. This surface configuration exactly corresponds to the charge modifications indicated in Fig. 3.7c, and to the micro-pyramid structure proposed more recently by Wolf (1992). Larger facets could also be imagined (Onishi *et al.*, 1987). In the general case, a precise estimation of the total energy has to take into account the lowering of surface energy due to the creation of dense facets, and the increase of energy associated with corner and edge atoms and with the effective increase in the surface area: this competition fixes the size of the facets.

Electronic structure

The electronic structure of polar surfaces does not follow the trends discussed at the beginning of the chapter for non-polar surfaces, primarily for two reasons. The first one comes from the number of broken bonds at the surface which may be large in some cases. For example, on the rocksalt{111} faces, three in every six anion–cation bonds are broken. But the major reason is linked to the behaviour of the electrostatic potential discussed above, which induces large charge redistributions. In some cases, these charge redistributions may provide the compensation of the macroscopic electric field required to stabilize the surface. The electronic structure of three polar oxide surfaces has been calculated: MgO{111}, ZnO(0001) and (000$\bar{1}$) and the weakly polar SrTiO$_3${001} face.

MgO{111} Tsukada and Hoshino (1982), using the DV-Xα method, have studied the electronic structure of a MgO cluster, with 26 atoms

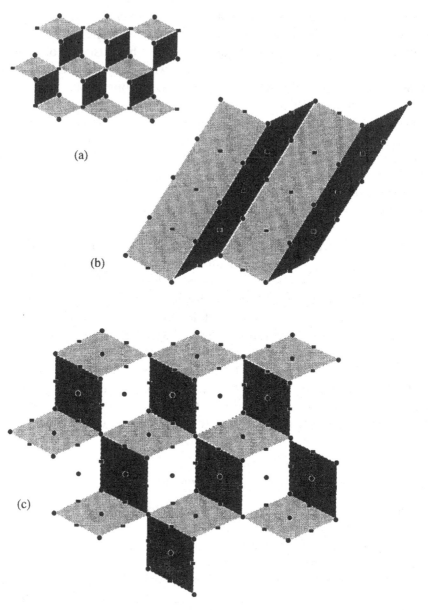

Fig. 3.8. Non-stoichiometric reconstructions of a rocksalt{111} face. (a) The stoichiometric surface; (b) every other atom is missing in the surface plane; (c) three out of four atoms are missing in the surface plane and one out of four in the underlying layer. Notice the {100} and {110} facets in (b) and the {100} facets in (c).

and four layers, oriented along the (111) direction. They found that the surface density of states is strongly modified compared to that of the bulk. The conduction band of the magnesium outer layer is shifted towards lower energies and the valence band of the oxygen outer layer is shifted towards higher energies, in agreement with the electrostatic potential variations represented on Fig. 3.7a. The surface bands overlap, leading to a metallization of the surfaces. The Fermi level lies at a position such that there is a large electron transfer from the surface oxygen valence band to the surface magnesium conduction band. A decrease σ' of the surface charges result. Despite important size effects, the authors find that σ' is of the order of $\sigma/2$. Similar results were obtained more recently in a slab geometry (Goniakowski, 1994).

Gibson *et al.* (1992) have also calculated the band structure of this surface in a slab geometry, by the total energy Harris–Foulkes functional method. They obtain a different behaviour on both surfaces: the magnesium surface becomes metallic as in the cluster calculation, while the oxygen surface remains insulating and consequently slightly more stable than the magnesium surface. The authors do not discuss the electrostatic effects.

Polar ZnO surfaces Tsukada *et al.* (1981) have calculated the electronic structure of small clusters modelling the ZnO(0001) and (000$\bar{1}$) faces, using the DV-Xα method. They find a decrease of the surface charges approximately equal to 25%, in agreement with the macroscopic arguments ($\sigma' = \sigma R/(R + R')$ and $R' = 3R$), despite the small size of the clusters. No metallization of the surfaces occurs. The authors argue that an increase of covalency of the surface anion–cation bond, resulting from the shifts of the atomic levels (Section 3.2), can provide the charge variation required for the surface stabilization, without an electron transfer from one side of the slab to the other, at variance with MgO{111}. They assign this difference to the larger covalent character of ZnO compared with MgO.

SrTiO$_3$(001) A calculation of the electronic structure of SrTiO$_3$(001) in a slab geometry, with eight atomic layers, has recently been performed, using the self-consistent tight-binding method (Goniakowski, 1994). The charge redistribution in the surface layers is found to fulfill the requirement $\sigma' = \sigma/2$. The bond-breaking at the surface provides the charge variation necessary to balance the macroscopic electric field. No metallization of the surfaces occurs.

Discussion The surface charge redistribution in the three oxides thus seems to have a different orgin. One can rationalize the three behaviours with the following qualitative arguments, which start from an ionic picture and treat the covalent effects as a small perturbation ($Z_A \beta^2 \ll (\epsilon_C - \epsilon_A)^2$).

Within this approximation, the electron number on each species depends linearly upon its local coordination. For example, in MgO, in the bulk ($Z_A = 6$), this leads to $Q_{Mg} = 2 - 6x$ and $Q_O = -2 + 6x$, with x the electron transfer per Mg–O bond. In ZnO, similarly, $Q_{Zn} = 2 - 4x$ and $Q_O = -2 + 4x$ ($Z_A = 4$), and in SrTiO$_3$, $Q_{Ti} = 4 - 6x$; $Q_{Sr} = 2 - 12y$ and $Q_O = -2 + 2x + 4y$, with x and y the electron transfer per Ti–O and Sr–O bond, respectively (each titanium has six oxygen neighbours and each strontium has twelve oxygen neighbours). The net charge in the total unit cell is zero in each case and, along the polar direction, the unit cells have a charge equal to $Q_v = \pm (2 - 6x)$, $Q_v = \pm (2 - 4x)$ and $+ Q_v = 2x - 8y$, respectively. In SrTiO$_3$, since titaniums are more electropositive than strontiums, x is larger than $4y$ and the SrO plane is positively charged.

On the SrO-terminated {100} face, the strontiums have eight neighbouring oxygens, while the oxygens have one neighbouring titanium and four neighbouring strontiums. According to the same assumption of charge-transfer additivity, and denoting x' and y' the charge transfers per Ti–O and Sr–O bond in the vicinity of the surfaces, one finds $Q_{Sr} = 2 - 8y'$ and $Q_O = -2 + x' + 4y'$, i.e. a total charge per SrO unit cell equal to: $Q_s = x' - 4y'$. Similarly, on the TiO$_2$-terminated {100} face, the titaniums have five neighbouring oxygens and the oxygens have two neighbouring titaniums and two neighbouring strontiums, which leads to $Q_{Ti} = 4 - 5x'$ and $Q_O = -2 + 2x' + 2y'$, i.e. a net charge equal and opposite to Q_s.

The bond-breaking process on this surface thus modifies the charges in such a way that the requirement $\sigma' = \sigma/2$ for the surface stabilization may be fulfilled without noticeable changes in the bond covalency ($x = x'$ and $y = y'$). In other words, contrary to the case of non-polar surfaces, the charge modification induced by bond-breaking is not screened; this absence of screening helps to decrease the electrostatic energy and to balance the macroscopic electric field.

If we apply the same model to the polar ZnO surfaces, on which the atoms are three-fold coordinated, the surface charges are equal to $Q_{Zn} = 2 - 3x'$ and $Q_O = -2 + 3x'$. The stabilization of the surface requires that the surface charge is equal to 3/4 the bulk one, i.e. $x' = x + 1/6$. This charge variation may be achieved by a mere shift of the atomic energy levels, because ZnO has a large part of covalent character. There is no need for metallization of the surfaces.

The same is not true on the MgO(111) surface. Assuming that the surface remains insulating, the surface charges read: $Q_{Mg} = 2 - 3x'$ and $Q_O = -2 + 3x'$. The condition $\sigma' = \sigma/2$ for the surface stabilization requires that $x' = x + 1/3$. This is a very large charge variation and, before it is achieved, the surface bands overlap leading to metallization of the surface.

Synthesis

To summarize, on insulating oxide polar surfaces, a reduction of the ionic charge density $\sigma' = \sigma R/(R + R')$ is required on the outer layers to cancel the macroscopic electric field, and stabilize the system. It may be induced by non-stoichiometry in the surface layers, by faceting or by charged impurities introduced during surface preparation. Systematic chemical analysis and measurements of the surface stoichiometry should be able to point out which mechanism takes place in a given case.

In mixed iono–covalent systems, the compensating charge may also result from an electron redistribution, due either to the bond-breaking process, to the change of ionicity of the anion–cation bond, or to a metallization of the surfaces inducing an electron transfer from one side of the sample to the other. Whether the system becomes metallic or not depends upon the specific self-consistent relationship between the charge and the surface Madelung potential and upon the importance of the charge redistribution which is required: $MgO(111)$ and $ZnO(0001)$ exemplify these two possibilities.

In ternary or more complex partly covalent oxides, some surfaces have a weak polarity, although they are considered to be non-polar according to Tasker's classification. However, the charge compensation on these weakly polar surfaces is more easily achieved than on polar surfaces, since it only involves a bond-breaking mechanism.

While several calculations have proved that a charge density $\sigma' = \sigma/2$ may appear at a surface when the electronic degrees of freedom are taken into account, it should be noted that no one has found a dipole density $\sigma''R$ with $\sigma'' = \sigma/4$. It is likely that, in actual systems with a discrete atomic structure, the oscillating macroscopic potential ($G_{\parallel} = 0$) (Fig. 3.7b) is strongly damped by microscopic non-uniform contributions ($G_{\parallel} \neq 0$ terms).

3.4 Non-stoichiometric surfaces

Defects on oxide surfaces are often produced when cutting a crystal, cleaning it by ionic bombardment, and annealing it at high temperature. It is very difficult to prepare a stoichiometric and structurally perfect surface. The surface treatments favour the creation of geometrical defects, such as steps, or stoichiometry defects such as oxygen or cation vacancies. Many experiments in the past have been performed, unwittingly, on imperfect surfaces and some of their results, for example those concerning the presence of intrinsic surface states, are questionable. More recent studies have been performed with a more rigorous control of the surface stoichiometry and are thus free of these limitations.

The interest in surface defects on oxides also relies on considerations of reactivity, since steps or vacancies are known to be active centres for chemical reactions. For example, in photo-electrochemical cells, anodes made of reduced TiO_2 or $SrTiO_3$ present a larger activity in the decomposition of water, than their stoichiometric counterparts, which was assigned to the presence of Ti^{3+} ions (Lo et al., 1978; Lo and Somorjai, 1978). The reduction of ReO_3 surfaces (Tsuda and Fujimori, 1981) and of $SnO_2(110)$ surfaces (Gercher and Cox, 1995) induces similar effects. Many structural and electronic studies have thus been devoted to the analysis of defects, and more particularly of oxygen vacancies.

Characteristics of the electronic structure

On MgO{100}, an electronic excitation has been recorded at 2.3 eV, i.e. at an energy much lower than the bulk gap width (7.8 eV). This transition was assigned to the presence of a surface F centre, associated with an oxygen vacancy, by Henrich et al. (1980) and by Henrich and Kurtz (1981b), despite a disagreement with theoretical predictions by Sharma and Stoneham (1976) and Kassim et al. (1978). Underhill and Gallon (1982), on the other hand, assigned it to a magnesium vacancy, inducing a surface V centre. Tsukada et al. (1983) predicted that surfaces containing oxygen vacancies should display electronic transitions at 5 eV, and they associated the 2 eV transition to the presence of steps. Kantorovich et al. (1995) have performed an ab initio study of defects, such as steps and oxygen vacancies on MgO(100). Castanier and Noguera (1995a) have considered the electronic features induced by various densities of oxygen vacancies on MgO(100).

On TiO_2{110} surfaces presenting a small amount of oxygen vacancies, Henrich et al. (1976) found, by ultra-violet photoemission, an occupied state at about 0.8 eV below the bottom of the conduction band. When the sub-stoichiometry increases, this state broadens and finally overlaps the conduction band. Using the STM in the spectroscopic mode, Zhong et al. (1992) proved that the absence of bridging oxygens is associated with a partial occupation of Ti 3d orbitals. Göpel et al. (1984) analyzed the electronic characteristics associated with oxygen vacancies by XPS, Auger and electron energy loss spectroscopies. They showed that the cation core levels are raised while the oxygen core levels are lowered on reduced surfaces, that the intensity of non-bonding O 2p states in the valence band decreases while the intensity of bonding states increases, and they confirmed that there exists a transition at 0.8 eV in the electron energy loss spectrum. They proposed a model in which the oxygen vacancies induce the presence of an occupied state 0.3 eV below the bottom of the conduction band, i.e. about 0.8 eV below the first peak in the unoccupied part of the density of states. More recently, Zhang *et al.* (1991) have estimated the strength of anion–cation hybridization in the valence band by resonant photoemission. They showed that it increases on reduced surfaces. Even more recently, Röhrer *et al.* (1992) and Zhong *et al.* (1992) have obtained information on the local electronic structure of reduced surfaces, by means of scanning tunneling microscopy operating in the spectroscopic mode. Theoretical studies performed by Tsukada *et al.* (1983), Munnix and Schmeits (1985) and Wang and Xu (1989) agree on the existence of a state associated with an oxygen vacancy below the conduction band. Munnix and Schmeits (1985) have considered several locations for the vacancy with or without atomic relaxation of the neighbouring cations. Ramamoorthy *et al.* (1994) have modelled the relaxation and rumpling of the (110) surface without its rows of bridging oxygens. Although the $TiO_2(110)$ and $SnO_2(110)$ surface structures are similar, their electronic characteristics happen to be quite different in the presence of oxygen vacancies. Egdell *et al.* (1986) showed by photoemission that, while the gap state lies 1 eV below the Fermi level on $TiO_2(110)$, it is located close to the top of the valence band in $SnO_2(110)$. Manassidis *et al.* (1995) have calculated the electronic structure on the relaxed sub-stoichiometric $SnO_2(110)$ face and found a broad distribution of gap states.

The $\{100\}$ face of $SrTiO_3$ also presents low-lying states in the gap at 1 eV and 1.5 eV below the conduction band minimum, which were assigned to oxygen vacancies and to defects resulting from surface preparation (Henrich *et al.*, 1976; 1978; Hikita *et al.*, 1993; Hirata *et al.*, 1994a). Several numerical calculations also found these states (Ellialtioglu *et al.*, 1978; Tsukada *et al.*, 1983). Yet, in the non-self-consistent calculation by Toussaint *et al.* (1987), a gap state can be found only if the relaxation of the cation positions in the neighbourhood of the vacancy is taken into account. On this same face, Brookes *et al.* (1987) have analyzed the role of steps as catalytic centres for water dissociation. Finally, Lo and Somorjai (1978) have studied the electronic structure of oxygen vacancies on the polar $\{111\}$ face of $SrTiO_3$.

The sub-stoichiometric (001) face of V_2O_5 presents a gap state and a charge transfer towards the vanadiums (Zhang and Henrich, 1994).

An increase of covalent character of the Zr–O bond was found in $ZrO_2(100)$ surfaces presenting oxygen deficiency (Sanz *et al.*, 1994).

The electronic structure of non-stoichiometric α-Al_2O_3 surfaces has been studied by electron energy losses and photoemission (Gautier *et al.*, 1991b; 1994; Gillet and Ealet, 1992). The reduction of the (0001) face induces a gap narrowing – from 8.3 eV on the stoichiometric surface to 7 eV on the reconstructed $(\sqrt{31} \times \sqrt{31})R9°$ face. In the valence band density of states, weight is transferred from non-bonding to bonding states on the non-stoichiometric surfaces.

The electronic structure of non-stoichiometric α-quartz surfaces was studied by LEED, XANES and EELS (Bart *et al.*, 1992; 1994). Transitions at energies equal to 5.2 eV and 7.4 eV – lower than the bulk gap – have been observed. They have been related to models of vacancies: in the quartz lattice.

Discussion

The above-mentioned studies stress some systematic characteristics associated with oxygen vacancies:

1) on most surfaces, a localized state appears in the gap;
2) when this state is occupied, it pins the Fermi level;
3) a decrease of the neighbouring cation charge results, which shifts the core levels;
4) the density of states in the valence band is distorted. Non-bonding states transfer weight to bonding states at the bottom of the band;
5) a gap reduction may occur.

These characteristics find their origin in the decrease of the Madelung potential felt by the Z cations which surround the vacancy, when an oxygen leaves the surface. The cation orbitals are polarized and their effective levels are lowered in energy. This increases their hybridization

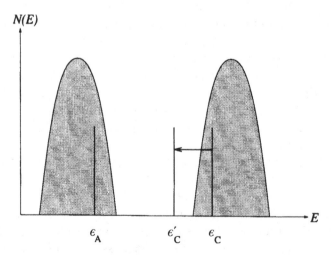

Fig. 3.9. Lowering of a cation effective level in the vicinity of a neutral oxygen vacancy and creation of a gap state.

with the remaining oxygens in their neighbourhood, and modifies the valence band shape, especially the strength of the bonding states at the bottom of the valence band. As a consequence of the cation level shift, a state generally appears in the gap of the band structure (Fig. 3.9). Its degeneracy depends upon the symmetry of the cation orbitals and upon Z. For example, for alkanine-earth oxides, it is Z-times degenerate. If the oxygen atom leaves the solid in a neutral state, two electrons come into the gap state. The latter is thus only partly filled and it pins the Fermi level. As a first approximation, the two electrons are shared between the Z cations, whose charge changes from $+ x$ to $+ (x - 2/Z)$, x depending upon the compound, the surface orientation, etc. If one is to obtain a good description of non-stoichiometric surfaces, a self-consistent approach is absolutely necessary.

The analytical model, presented in Section 3.1 to interpret the electronic properties of perfect surfaces, may be extended (Castanier and Noguera, 1995a) to describe non-stoichiometric surfaces). In particular the description of the *local* cation–anion hybridization, outlined in (3.2.6) to (3.2.8), allows one to take into account: 1) the loss of neighbours of the cations; 2) the lowering of the cation levels and the increase of hybridization with the remaining oxygens; and 3) the filling of the gap state, which are the three prominent features of the cation LDOS. For example on MgO(100), for a magnesium close to Z_{vac} vacancies, the cation electron number, summed over s and p orbitals, reads:

$$
N_C = \left(1 - \frac{(\epsilon_{Cs} - \epsilon_O)}{\sqrt{(\epsilon_{Cs} - \epsilon_O)^2 + 4(5 - Z_{vac})\beta_s^2}} \right)
$$
$$
+ 3 \left(1 - \frac{(\epsilon_{Cp} - \epsilon_O)}{\sqrt{(\epsilon_{Cp} - \epsilon_O)^2 + 4(5 - Z_{vac})\beta_p^2}} \right)
$$
$$
+ \frac{Z_{vac}}{4} \left(1 + \frac{(\epsilon_{Cs} - \epsilon_O)}{\sqrt{(\epsilon_{Cs} - \epsilon_O)^2 + 4(5 - Z_{vac})\beta_s^2}} \right) . \quad (3.4.1)
$$

This expression is obtained by integrating the cation local density of states up to the Fermi level, over the valence band and the filled part of the gap state. The first two terms correspond respectively to the s and p electron numbers, due to the filling of the valence band. On stoichiometric surfaces ($Z_{vac} = 0$) and with the appropriate effective atomic energies, they yield the magnesium electron number N_{C0} discussed in Section 3.1. These two terms vanish in the ionic limit where the $(\epsilon_{Ci} - \epsilon_O)$ are much larger than $4(5 - Z_{vac})\beta^2$. Compared to the stoichiometric surface, a rough balance between the reduction in the coordination number (from 5 to $5 - Z_{vac}$)

and the reduction in $(\epsilon_{Ci} - \epsilon_O)$ (due to the lowering of the cation levels) takes place. The third term in (3.4.1) comes from the filling of the gap state. It corresponds to the sharing of two electrons between the four surface magnesiums which surround the vacancy. These latter receive two electrons only in the ionic limit. Otherwise, a fraction of the electrons is transferred to the $(5 - Z_{vac})$ remaining oxygens.

When more vacancies are formed, cations in the surface layer may have non-equivalent environments and their local electronic structure depends upon the number of vacancies in their first coordination shell. Understanding the charge transfers, the position of the gap states and, more generally, the energetics of these processes is important to conclude whether vacancies attract each other, leading to a clustering of vacancies, or repel each other, leaving the possibility of ordered periodic structures in the surface layer. In contrast to the numerous experimental evidences of non-stoichiometric surface reconstructions on oxide surfaces, recalled in Chapter 2, no theoretical study has yet been advanced to describe the mechanisms responsible for vacancy ordering. Non-stoichiometric rutile surfaces have been studied, under the assumption of a vacancy distribution close to the one seen in experiments (Oliver *et al.*, 1994; Ramamoorthy *et al.*, 1994), but no research of the configuration of vacancies with the lowest energy has been undertaken. A first attempt to answer this question was performed on non-stoichiometric MgO(100) surfaces, leading to the conclusion that an effective attraction between vacancies exists due to Madelung interactions. Long-range order between vacancies does not provide any energy lowering on this surface at constant vacancy density (Castanier and Noguera, 1995b). This result is consistent with the observation of clusters of vacancies on MgO(100) (Nakamatsu *et al.*, 1989) and of a clustering of nickel atoms on reduced NiO(100) surfaces when the vacancy density exceeds 50% (Wulser *et al.*, 1992). The reasons for the different behaviour of vacancy–vacancy interactions on rutile and alumina surfaces remain unclear.

3.5 Conclusion

Peculiarities in the electronic structure of oxide surfaces are related to the strength of the electrostatic potential which acts on the surface atoms and to the number of broken anion–cation bonds. In a qualitative way, the more open the surface, the stronger the Madelung potential reduction and the less efficient the electron delocalization.

The reduction of the number of anion–cation bonds, which decreases the space available for electron delocalization on surfaces, leads to modifications in the band shape, to the appearance of surface states in the bulk gap and acts towards an increase of ionicity as far as the ionic charges are concerned.

The modification of the Madelung potential, on the other hand, induces important shifts of the surface atomic energy levels compared to those in the bulk: towards higher energies for the anion levels and towards lower energies for the cation levels. These shifts are direcly reflected in the gap width and may strongly modify the transport properties of surface layers or thin films. In addition, as will be proved in Chapter 6, the reactivity of surface sites depends, among other parameters, upon the energy position of the orbitals which take part in the chemical bond with the reactive species: a detailed understanding of the factors which determine these positions is thus important.

As far as charges are concerned, on non-polar surfaces the electrostatic effect approximately balances the reduction of hybridization. On polar surfaces, however, electrostatic effects are more drastic. Depending upon the compound, they may lead to surface instabilities, spontaneous desorption of atoms followed by non-stoichiometric reconstructions, or to surface metallization.

On sub-stoichiometric surfaces, electrostatic effects are responsible for the appearance of gap states, closely related to the positions of the effective levels of the cations which surround the vacancy. These gap states are generally partly filled.

Many questions related to the electronic and atomic structures of oxide surfaces remain unanswered. For example, the respective roles of bond-breaking and structural distortions in the interpretation of surface densities of states and gaps are not well elucidated. Similarly, for complex crystal structures, it is difficult to relate the observations to models, since one does not know precisely the surface terminations. More systematic studies of stoichiometry in surface layers, and of the electronic and atomic features of polar surfaces are needed to better disentangle the origin of the deep surface states which may result either from dangling bonds or from structural and stoichiometric defects. Finally, the modification of correlation effects on oxide surfaces represents a completely virgin field for future investigations.

4

Surface excitations

A semi-infinite crystal supports specific excitations which are not present in the bulk material. In oxides, these new modes result from the breaking of anion–cation bonds at the surface, whose effects on the electronic and atomic structure have been the subject of the two previous chapters. Here, we will describe the excitations associated with the atomic and electronic degrees of freedom. The phonons, which are the quantized modes of vibrations of the atoms, have small characteristic energies, of the order of a few tens of milli-electron volts (meV). The electron–hole pairs and plasmons, which are characteristic of the electronic degrees of freedom, have much higher energies, of the order of a few electron-volts. Due to the different time scales involved, a decoupling between these two types of excitations takes place (Born–Oppenheimer decoupling). The electrons are able to follow 'instantly' the atomic displacements, while the atoms have no time to move at the time scale of the electron delocalization and excitations.

4.1 Surface phonons

We have seen in Chapter 2 that bond-breaking on a surface greatly modifies the atomic energy levels. The same is true for the atomic vibrations around the equilibrium positions. In this section, we will review the main experimental and theoretical results concerning surface phonons on oxide surfaces.

Experimental and theoretical approaches

High-resolution electron energy loss (HREELS = high-resolution electron energy loss spectroscopy) experiments and inelastic scattering of helium atoms allow a determination of the surface vibration modes and of the phonon dispersion curves. The same cannot be achieved by inelastic neutron scattering, due to the high penetration depth of the neutrons in matter. In HREELS, a low-energy electron beam is impinged on a sample and the energy losses in the phonon range (0 to 100 meV) are recorded. The resolution is mainly determined by the energy dispersion of the incident beam. In the most sensitive set-ups, it may reach 1 meV. Low-energy helium atoms (10 meV) may also be used to sample the surface structure and atomic vibrations.

The surface phonons of rocksalt crystals have been thoroughly studied (Kress and de Wette, 1991). Among the insulating oxides, the frequencies of the optical surface modes – Fuchs and Kliewer phonons – have been determined in MgO(001) (Brusdeylins *et al.*, 1983; Thiry *et al.*, 1984; Liehr *et al.*, 1986; Cui *et al.*, 1990), CoO(001) (Nassir and Langell, 1994), ZnO($1\bar{1}00$) (Ibach, 1970; 1972), NiO(100) (Dalmai-Imelik *et al.*, 1977; Andersson and Davenport, 1978; Cox and William, 1985; Wulser and Langell, 1994), SrTiO$_3$(001) (Baden *et al.*, 1981; Cox *et al.*, 1983; Conard *et al.*, 1993), α-Al$_2$O$_3$(0001) (Liehr *et al.*, 1984; 1986), SiO$_2$(0001) (Thiry *et al.*, 1985) and TiO$_2$(110) (Kesmodel *et al.*, 1981). More recently the phonon dispersion of thin alumina films deposited on a ruthenium(001) substrate has been obtained (Frederick *et al.*, 1991a).

Theoretically, the elasticity theory of continuous media may be used to study the long-wavelength modes. To determine the microscopic modes, numerical approaches are necessary. Most of them have used Born's model to estimate the inter-atomic forces. The semi-infinite crystals are modelled by thin films, whose thickness must be larger than the attenuation length of the surface modes. The complete MgO(001) phonon spectrum has been calculated, neglecting (Chen *et al.*, 1977; Barnett and Bass, 1979) or taking into account (Lakshmi and de Wette, 1980) the surface relaxation. The same has been done for SrTiO$_3$(001) (Prade *et al.*, 1993).

Classification of the surface modes

The modes are classified according to the polarization of the atomic displacements: they are acoustic (A) when all atoms in a unit cell move in phase, or optic (O) when anions and cations within a given unit cell have displacements of opposite directions. The modes have a pure longitudinal (L) or transverse (T) character along high-symmetry lines in reciprocal space, when the atomic motions are parallel or perpendicular to the saggital plane – the plane which contains the normal to the surface and the wave vector. In order to discuss surface modes, it is important to know the bulk dispersion relation, and especially the location of the gaps in reciprocal space. For example, Fig. 4.1 shows the phonon dispersion curve for MgO(100) (Chen *et al.*, 1977). More generally, in rocksalt compounds, three gaps may be identified:

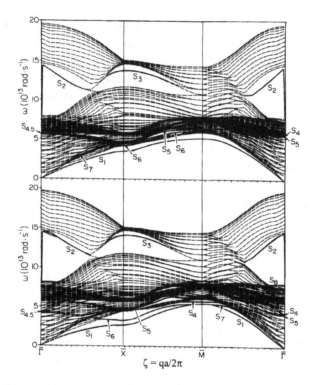

Fig. 4.1. Phonon dispersion curves in MgO(100) (according to Chen *et al.*, 1977). Hatched zones are the projection of the bulk modes. Surface modes S_n are indexed by n ($1 \leq n \leq 7$): the Rayleigh mode is S_1; the Fuchs and Kliewer modes have a frequency close to 12×10^{13} rad s^{-1}; S_3 is an example of a microscopic mode.

- a gap between the longitudinal acoustic (LA) and the transverse acoustic (TA) branches, which is located around $\overline{X}$,
- a gap between the longitudinal optic (LO) and transverse optic (TO) branches, which is open at $\overline{\Gamma}$, which closes at one-third of the distance between $\overline{\Gamma}$ and $\overline{X}$ and which re-opens at $\overline{X}$,
- an absolute gap between the transverse optic (TO) and longitudinal acoustic (LA) branches, present in the whole Brillouin zone.

Depending upon their penetration into the deeper layers, the surface modes are called macroscopic or microscopic modes:

- The macroscopic modes have an attenuation length which varies as the inverse of the normal component of their wave vector. They penetrate deeply inside the crystal. On this length scale, the precise atomic structure is unimportant and the elasticity theory of continuous media or dielectric theories may be used. Depending upon whether they are acoustic or optic, one distinguishes the Rayleigh mode (1885) and the Fuchs and Kliewer modes (1965).

- The microscopic modes have a penetration depth which is of the order of a few inter-plane spacings. Their description requires a precise account of the atomic structure. They are generally located in the gaps of the bulk phonon spectrum. However, at some positions in reciprocal space, they may become degenerate with the bulk modes, thus transforming into surface resonances. Lucas (1968) was one of the first authors to predict their existence in ionic crystals.

Rayleigh mode

The first mention of surface phonons is due to Lord Rayleigh (1885), who predicted the existence of a surface acoustic mode with a sound velocity lower than in the bulk. He proved this result, using elasticity theory, by representing the semi-infinite solid by a continuous and isotropic medium (Landau and Lifchitz, 1967). Considering an infinitesimal volume element, he wrote a Fourier component of its displacement $\vec{u}(\vec{q}, \omega)$, in the following form:

$$\vec{u}(\vec{q}, \omega) = \vec{a} \exp\left[i\left(q_\parallel x - \omega t\right)\right] \exp(-\kappa z), \qquad (4.1.1)$$

with κ a positive quantity given by (c the light velocity):

$$\kappa = \sqrt{q_\parallel^2 - \frac{\omega^2}{c^2}}. \qquad (4.1.2)$$

The $\vec{u}$ vector has transverse ($\vec{\nabla} \bullet \vec{u}_t = 0$) and longitudinal ($\vec{\nabla} \times \vec{u}_\ell = 0$) components, associated with sound velocities c_t and c_ℓ, respectively. Boundary conditions require that all the volume elements are in equilibrium, especially at the surface ($z = 0$):

$$\sigma_{iz} dA_z = 0. \qquad (4.1.3)$$

The tensor of stresses σ_{ij} is related to the tensor of deformations, to lowest order, by Hooke's law. The relevant elastic constants are the Young modulus Y and the Poisson's coefficient P.

A surface mode, with a polarization parallel to the saggital plane xz, obeys the boundary conditions. The longitudinal and transverse components of the $\vec{u}$ vector fulfil the following relations:

$$iq_\parallel u_{tx} + \kappa_t u_{tz} = 0$$
$$iq_\parallel u_{\ell z} + \kappa_\ell u_{\ell x} = 0. \qquad (4.1.4)$$

By writing down that $\sigma_{xz} = \sigma_{zz} = 0$, i.e.:

$$\frac{\partial u_x}{\partial z} + \frac{\partial u_z}{\partial x} = 0$$
$$c_\ell^2 \frac{\partial u_z}{\partial z} + (c_\ell^2 - 2c_t^2)\frac{\partial u_x}{\partial x} = 0, \qquad (4.1.5)$$

the surface mode is found to be an acoustic mode: $\omega = c_R q_{\parallel}$ with a sound
velocity equal to $c_R = \xi c_t$. The ξ constant is the solution of the equation:

$$\xi^6 - 8\xi^4 + 8\xi^2 \left(3 - 2\frac{c_t^2}{c_\ell^2}\right) - 16\left(1 - \frac{c_t^2}{c_\ell^2}\right) = 0 . \qquad (4.1.6)$$

It depends upon the ratio $c_t/c_\ell = \sqrt{(1 - 2P)/2(1 - P)}$. Only one root of
(4.1.6) fulfils the condition $\xi < 1$ required for κ_t and κ_ℓ to be real. Since
c_t/c_ℓ is larger than zero and smaller than $1/\sqrt{2}$ ($0 \le P \le 1/2$), ξ belongs
to the interval [0.874, 0.955]. The sound velocity of the Rayleigh mode is
thus lower than the lowest sound velocity in the bulk. Close to the zone
centre, the mode is located below the bulk acoustic modes (S_1 mode in
Fig. 4.1).

Good agreement was found for the dispersion of the Rayleigh mode
in MgO(100), between the results of helium scattering experiments (Jung
et al., 1991) and calculations using Born's model and neglecting surface
distortions (Chen *et al.*, 1977). This was said to support the idea that relaxation, rumpling and charge redistributions on MgO(100) are negligible.
Actually, the dispersion of the Rayleigh mode is likely to be insensitive to
the details of the atomic structure at the surface due to its long penetration
length.

Fuchs and Kliewer modes

Fuchs and Kliewer (1965) have predicted the existence of macroscopic
surface optic modes in ionic crystals. We give here a simplified derivation
of their result, based on the formalism of the dielectric constant. In the
phonon frequency range, the bulk dielectric constant $\epsilon(\omega)$ approximately
varies with ω as:

$$\epsilon(\omega) = \epsilon_\infty + (\epsilon_0 - \epsilon_\infty)\frac{\omega_{TO}^2}{(\omega_{TO}^2 - \omega^2)} , \qquad (4.1.7)$$

or, according to the Lyddane–Sachs–Teller relation, as:

$$\epsilon(\omega) = \epsilon_\infty\frac{\omega_{LO}^2 - \omega^2}{\omega_{TO}^2 - \omega^2} . \qquad (4.1.8)$$

At a surface, new phonon modes are associated with the roots of the
equation:

$$1 + \epsilon(\omega_s) = 0 , \qquad (4.1.9)$$

which results from a matching procedure at the surface. A similar equation
will be used in Section 4.3 to derive the frequency of surface plasmons.
Taking into account the expression (4.1.8) for the bulk dielectric constant,

ω_s is equal to:

$$\omega_s = \omega_{TO}\sqrt{\frac{\epsilon_0 + 1}{\epsilon_\infty + 1}} . \qquad (4.1.10)$$

It is larger than ω_{TO} and smaller than ω_{LO}. Lucas and Vigneron (1984) and Lambin *et al.* (1985) have generalized this dielectric approach to anisotropic materials and to interfaces.

More precisely, Fuchs and Kliewer have shown that there exist two modes ω_{FK+} and ω_{FK-} whose frequencies are close to ω_{TO} and ω_{LO} at the zone centre, and which become degenerate as the product $q_\parallel \times a$, where a is the layer thickness, increases. Their common value is intermediate between ω_{TO} and ω_{LO}. At the zone centre, the surface mode is uniform in the whole slab. In MgO, for example, it has a frequency close to 1.2×10^{14} rad s^{-1}.

Microscopic modes

It is difficult to build a universal model which reasonably describes the microscopic surface modes in real materials, because they are sensitive to the actual crystal structure and surface orientation. Nevertheless, in order to exemplify some of their characteristics, we now present a calculation of the vibrations in a semi-infinite linear chain of alternating anions and cations. The atoms, with respective masses M_A and M_C (reduced mass μ) are equidistant and assumed to be linked by springs of stiffness τ. The displacements of the anions and cations, with respect to their equilibrium positions, in the cell n, are respectively denoted u_n and v_n. In the surface cell ($n = 1$), an anion is in contact with vacuum. The equations for motion in a cell $n \neq 1$ are:

$$M_A\frac{d^2 u_n}{dt^2} = -\tau\,(2u_n - v_n - v_{n-1})$$

$$M_C\frac{d^2 v_n}{dt^2} = -\tau\,(2v_n - u_n - u_{n+1}) , \qquad (4.1.11)$$

and in the surface cell ($n = 1$):

$$M_A\frac{d^2 u_1}{dt^2} = -\tau\,(u_1 - v_1)$$

$$M_C\frac{d^2 v_1}{dt^2} = -\tau\,(2v_1 - u_1 - u_2) . \qquad (4.1.12)$$

For a given Bloch wave vector, the bulk modes $u_n = u\exp(ikna)\exp(i\omega t)$ and $v_n = v\exp(ikna)\exp(i\omega t)$ satisfy (4.1.11). Their frequencies are equal to:

$$\omega_k^2 = \frac{\tau}{\mu} \pm \frac{\tau}{M_A M_C}\sqrt{(M_A - M_C)^2 + 4M_A M_C \cos^2(ka/2)} . \qquad (4.1.13)$$

The mode associated with the '−' sign is an acoustic mode. Its frequency vanishes at the zone centre. The second mode is an optic mode with a frequency at $k = 0$ equal to $2\tau/\mu$.

A surface mode is a singular solution of the bulk equations of motion, which obeys the surface boundary conditions. Its Bloch wave vector may be complex. In the case of the alternating linear chain, combining the two systems of equations leads to:

$$\exp(ika) = -\frac{M_A}{M_C} , \qquad (4.1.14)$$

which defines the penetration depth. The frequency:

$$\omega_s^2 = \frac{\tau}{\mu} , \qquad (4.1.15)$$

is intermediate between the acoustic and optic frequencies. Using a calculation of this type applied to thin layers, Lucas (1968) has obtained the frequencies of the microscopic modes at a NaCl(100) surface.

Soft phonons; reconstructions; structural phase transitions

In the process of surface formation, there may exist positions $\vec{k}$ in reciprocal space where the frequency of a vibration is so deeply modified, as a consequence of bond-breaking, that it vanishes. The mode becomes unstable and one speaks of a soft phonon. The atoms are no longer pulled back to their equilibrium positions. A new periodicity, characterized by the wave vector $\vec{k}$, shows up in low-energy electron or helium atom diffraction experiments. The mechanism of phonon softening was suggested to explain some reconstructions observed on metal surfaces (Blandin *et al.*, 1973; Fasolino *et al.*, 1980; Pick, 1990). When a true structural phase transition occurs at a temperature T_c, the frequency of the soft mode decreases as the temperature decreases in the fluctuative regime ($T > T_c$). Below the critical temperature T_c, superstructure satellites appear. Such phenomena were, for example, observed by Ernst *et al.* (1987) on W(001), by diffraction of helium atoms. They are also present in quasi-one-dimensional conductors which display a Peierls' transition. One should nevertheless remember that there exist structural phase transitions that are not driven by soft phonons.

On SrTiO₃(001), Prade *et al.* (1993) have calculated the relaxation strength, the origin of the reconstruction and the vibration dynamics, using Born's model, with an account of the ionic polarization by the shell model. They have analyzed the manifestation of the anti-ferrodisplacive phase transition at the surface, a transition which involves rotations of TiO₆ octahedra and which induces a doubling of the lattice parameters. Introducing in their model force constants which vary with temperature, they have proved that at $\overline{M}$ in two-dimensional reciprocal space, a phonon mode softens, in agreement with inelastic neutron experiment results (Shirane and Yamada, 1969). The transition is predicted to take place at temperatures T_{s1} and T_{s2} which are different on the TiO_2 and SrO non-equivalent surfaces, but which are both larger than the bulk critical temperature ($T_c = 105$ K). The surface reconstruction is thus expected to occur at a temperature higher than the bulk critical temperature. There is no exprimental proof of this prediction. The soft phonon mode was detected in the fluctuative regime by electron energy loss spectroscopy (Conard *et al.*, 1993).

4.2 Bulk electronic excitations

Electronic excitations are produced when a solid is submitted to an applied electric perturbation of the proper frequency. Their energy spectrum may be studied by inelastic scattering experiments. Their characteristic length determines the spatial distribution of the charge induced by the perturbation.

Unlike the case of metals and semi-conductors, surface and bulk screening effects in insulators have been little studied. In this section, we will review the general properties of the dielectric constant – its small wave vector and low-frequency limits – and we will put a particular emphasis on local field effects.

General features

Screening effects are one of the most important manifestations of the existence of electron–electron interactions in solids. To discuss them, we will first consider a spatially homogeneous system, in which the response at a position $\vec{r}$ to an electric perturbation localized at $\vec{r}_0$ only depends upon $\vec{r} - \vec{r}_0$. This is true, for example, in an homogeneous interacting electron gas. The concept of the dielectric constant refers to the response to a weak perturbation. The relationship between the modification of the charge density $\delta\rho(\vec{q}, \omega)$ and the electrostatic potential $V(\vec{q}, \omega)$, is linear in this case, which is the range of validity of the linear response theory. It is then possible to define the electronic susceptibility $\chi(\vec{q}, \omega)$, expressed in double Fourier space:

$$\delta\rho(\vec{q}, \omega) = \chi(\vec{q}, \omega)V(\vec{q}, \omega) . \tag{4.2.1}$$

In the Random Phase Approximation (RPA), the potential $V(\vec{q}, \omega)$ to be used in (4.2.1) has two contributions: the applied external potential $V_{ext}(\vec{q}, \omega)$ which perturbs the system and the induced potential due to the change in the electronic wave functions. By definition, the dielectric constant $\epsilon(\vec{q}, \omega)$ relates the total electrostatic potential $V(\vec{q}, \omega)$ to $V_{\text{ext}}(\vec{q}, \omega)$:

$$V(\vec{q}, \omega) = \frac{V_{\text{ext}}(\vec{q}, \omega)}{\epsilon(\vec{q}, \omega)} . \qquad (4.2.2)$$

The electronic susceptibility $\chi(\vec{q}, \omega)$ depends upon the electronic structure of the system: its eigenstates $|\vec{k}\lambda\rangle$ and its eigenenergies $E(\vec{k}, \lambda)$ (λ, the band index), and upon the Fermi–Dirac function $f_0(\vec{k}, \lambda)$:

$$\chi(\vec{q}, \omega) = - \sum_{\vec{k}, \lambda, \lambda'} \frac{\left|\langle \vec{k}, \lambda | e^{i\vec{q}\vec{r}} | \vec{k} + \vec{q}, \lambda'\rangle\right|^2 \left\{f_0(\vec{k}, \lambda) - f_0(\vec{k} + \vec{q}, \lambda')\right\}}{E(\vec{k} + \vec{q}, \lambda') - E(\vec{k}, \lambda) - \hbar\omega} . \qquad (4.2.3)$$

It takes into account all intra-band and inter-band transitions between filled and empty states, and thus characterizes the total polarizability of the system. The link between the electronic susceptibility and the dielectric constant is the following ($-e$ the electron charge):

$$\epsilon(\vec{q}, \omega) = 1 - \frac{4\pi e^2}{q^2} \chi(\vec{q}, \omega) . \qquad (4.2.4)$$

Equations (4.2.3) and (4.2.4) relate the screening properties of a given system to its electronic structure.

Static $\omega = 0$ and macroscopic $q \to 0$ limits

We will first consider processes which are static on the time scale of electronic transitions ($\omega = 0$) but very fast on the time scale of atomic vibrations ($\omega = \infty$). The macroscopic limit is obtained for infinitely small q values. The main difference between a metal and an insulator lies in the denominator of $\chi(\vec{q})$. In insulators or semi-conductors, the smallest energy difference $E(\vec{k} + \vec{q}) - E(\vec{k})$ between filled and empty levels is equal to the gap width Δ, while in metals, $E(\vec{k} + \vec{q}) - E(\vec{k})$ may vanish. Using the relationship (Ziman, 1964):

$$\sum_{\vec{k}} \left[E(\vec{k}) - E(\vec{k} + \vec{q})\right] \left|\langle\vec{k}|e^{i\vec{q}\vec{r}}|\vec{k} + \vec{q}\rangle\right|^2 = \frac{\hbar^2 q^2}{2m} , \qquad (4.2.5)$$

obtained from the calculation of the double commutator $[[H, e^{i\vec{q}\vec{r}}], e^{-i\vec{q}\vec{r}}]$ and using the approximation $E(\vec{k} + \vec{q}) - E(\vec{k}) \approx E_g$, one finds that the susceptibility is proportional to q^2 when q is small, and that the dielectric constant reaches a finite value at $\vec{q} = \vec{0}$:

$$\epsilon(\vec{q} = \vec{0}, \omega = 0) \approx 1 + \frac{4\pi n e^2 \hbar^2}{m E_g^2} , \qquad (4.2.6)$$

which may be rewritten in the following form:

$$\epsilon(\vec{q} = \vec{0}, \omega = 0) \approx 1 + \left(\frac{\hbar\omega_{\mathrm{p}}}{E_{\mathrm{g}}}\right)^2. \tag{4.2.7}$$

In the literature, $\epsilon(\vec{q} = \vec{0}, \omega = 0)$ is called the optical dielectric constant and denoted ϵ_∞ because of the short time scale it involves compared to the phonon periods. ϵ_∞ is small in wide-gap insulators. Although the energy E_{g} in the denominator of (4.2.7) is not the gap Δ but rather the energy separation between the centres of gravity of the valence and conduction bands, ϵ_∞ qualitatively increases as Δ gets smaller and diverges in the metallic limit ($\Delta = 0$). Metals thus behave as perfect electrostatic mirrors. Typically, ϵ_∞ is equal to 3 for MgO, 6.8 and 8.4 for TiO_2 and 2.4 for SiO_2 (Gray, 1963). Advanced electronic structure calculations account reasonably well for the screening processes and the values of the dielectric constants. However, they have mostly been used in semi-conductors and metals. To the author's knowledge, among the insulators, only LiCl (Hybertsen and Louie, 1987a; 1987b), NiO (Aryasetiawan *et al.*, 1994) and $LiNbO_3$ (Ching *et al.*, 1994) have been modelled numerically.

The plasma frequency ω_{p} appears in the numerator of $\epsilon(\vec{q} = \vec{0}, \omega = 0)$. It depends on the electron density n in the valence band, in the following way:

$$\omega_{\mathrm{p}} = \sqrt{\frac{4\pi n e^2}{m}}, \tag{4.2.8}$$

as in metals. The values of ω_{p} are similar in metals and insulators. Yet, unlike metals, we will see in the following that ω_{p} is not the frequency at which $\epsilon(0, \omega)$ vanishes in insulators.

Static limit $\omega = 0$; spatial dependence of screening processes

In a free electron gas, screening processes have been the subject of detailed investigations, using either the Thomas–Fermi or the Lindhard dielectric constants. The charge induced by an external point charge remains localized in the close vicinity of the perturbing charge, i.e. within a distance of the order of the Thomas–Fermi screening length, which is a fraction of an ångström. There also exist small long-range oscillations of the electron density, called the Friedel oscillations, characterized by a wave vector equal to twice the Fermi wave vector.

In semi-conductors, screening effects have been described in a much less detailed way, using an assumption of spatial homogeneity and isotropy of space, and a simplified account of the electronic structure (Penn, 1962; Bechstedt and Enderlein, 1979; Levine and Louie, 1982; Cappellini *et al.*,

1993). For example, Cappellini *et al.* (1993) have proposed for $\epsilon(q, \omega = 0)$ the following expression (ζ an empirical parameter):

$$\epsilon(q, \omega = 0) = 1 + \frac{1}{1/(\epsilon_\infty - 1) + \zeta(q/q_{\text{TF}})^2 + \hbar^2 q^4/4m^2\omega_p^2} . \quad (4.2.9)$$

In the limit of small q, $\epsilon(q, \omega = 0)$ has the asymptotic form:

$$\epsilon(q \to 0, \omega = 0) = 1 + \frac{\epsilon_\infty - 1}{1 + \zeta q^2(\epsilon_\infty - 1)/q_{\text{TF}}^2} , \quad (4.2.10)$$

At $q = 0$, $\epsilon(q, \omega = 0)$ is equal to the optical dielectric constant ϵ_∞. Equation (4.2.9) gives the contribution of free electrons for large q:

$$\epsilon(q \to \infty, \omega = 0) = 1 + \frac{\hbar^2\omega_p^2}{(\hbar^2 q^2/2m)^2} . \quad (4.2.11)$$

By comparing their model to experimental results obtained in Si, Ge, GaAs, and ZnSe, the authors have found that the empirical parameter ζ is roughly constant in the series: $\zeta \approx 1.56$. They deduce, from (4.2.9), the coulomb hole expression as a function of the local density $\rho(\vec{r})$, the Fermi k_F and Thomas–Fermi q_{TF} wave vectors.

When a point charge Q is added at $\vec{r} = \vec{0}$, to a system described by the dielectric constant given in (4.2.9), the screened potential $V(r)$ varies with r in the following way:

$$V(r) = \frac{Q}{\epsilon_\infty r} + Q\left(1 - \frac{1}{\epsilon_\infty}\right)\frac{\exp(-q_{\text{TF}}r)}{r} . \quad (4.2.12)$$

This expression is obtained by setting ζ equal to 1 and forgetting the q^4 terms in (4.2.9). At long distances, there remains a long-range coulomb potential, due to an effective charge Q/ϵ_∞, in agreement with macrocopic electrostatic arguments. At shorter distances, the second term in the potential represents the contribution of an induced charge equal to $-Q(1 - 1/\epsilon_\infty)$, localized in a sphere of radius equal to the Thomas–Fermi screening length $1/q_{\text{TF}}$. Compared to a metal, a semi-conductor is unable to provide a complete screening to an external perturbation. The induced charge is smaller than the perturbing charge in absolute value.

Dynamic screening $\omega \neq 0$; macroscopic limit $q \to 0$

The dynamic response of the electrons to a time-varying ($\omega \neq 0$) homogeneous ($q \to 0$) perturbation is determined by $\epsilon(q \to 0, \omega)$. The elementary electronic excitations of the systems are of two types: the inter-band transitions, associated with the creation of electron–hole pairs, which correspond to the poles of $\epsilon(q \to 0, \omega)$ and the collective plasmon excitations. The inter-band transitions with the lowest energies are produced when an electron is excited from the valence band into the conduction band. In insulators and semi-conductors, they require an energy larger than the gap width Δ. Nevertheless, the threshold generally occurs at an energy lower than Δ, due to the attractive coulomb interaction between the electron which has been excited into the conduction band and the hole which is left behind in the valence band. Such an excitation with an energy lower than Δ is called an exciton. When the valence and conduction bands are flat, and when the optical dielectric constant is small, the electron and the hole remain localized on the same atom and move coherently in the solid. They form a Frenkel exciton, whose binding energy is equal to a non-negligible fraction of the gap width. In the opposite limit, the exciton is more diffuse and the electron–hole pair may be compared to an hydrogenoid system with quantized energy levels. It is a Wannier exciton (Elliott, 1957; Cho, 1979).

In oxides, excitonic transitions have been invoked to interpret the results of optical absorption and reflectivity experiments. They have also been invoked to explain discrepancies between experimental spectra and calculated joined densities of states, in quartz SiO_2 (Pantelides and Harrison, 1976; Chelikowsky and Schlüter, 1977; Gupta, 1985), in $SrTiO_3$ (Xu *et al.*, 1990), in MgO, Al_2O_3 and $MgAl_2O_4$ (Cohen *et al.*, 1967; Bortz and French, 1989; Xu and Ching, 1991). In most cases, the absorption intensity presents an enhancement at low energy, in the interband transition continuum, rather than a well-defined peak located below this continuum.

Collective excitations, called plasmons, are associated with a divergence of the screening response. Their energies $\hbar\omega_b$, in the bulk, are such that $\epsilon(q, \omega_b)$ vanishes. To estimate $\hbar\omega_b$, the susceptibility $\chi(\vec{q}, \omega)$ is rewritten in the following form:

$$\chi(\vec{q}, \omega) = -\sum_{\vec{k}, \lambda, \lambda'} \frac{|\langle \vec{k}, \lambda | e^{i\vec{q}\vec{r}} | \vec{k} + \vec{q}, \lambda' \rangle|^2 f_0(\vec{k}, \lambda)}{E(\vec{k} + \vec{q}, \lambda') - E(\vec{k}, \lambda) - \hbar\omega}$$
$$+ \sum_{\vec{k}, \lambda, \lambda'} \frac{|\langle \vec{k}, \lambda | e^{i\vec{q}\vec{r}} | \vec{k} + \vec{q}, \lambda' \rangle|^2 f_0(\vec{k}, \lambda)}{E(\vec{k}, \lambda) - E(\vec{k} - \vec{q}, \lambda') - \hbar\omega} \ . \tag{4.2.13}$$

Using (4.2.5) and making the assumption $E(\vec{k} + \vec{q}) - E(\vec{k}) \approx E_g$, one finds successively:

$$\epsilon(q = 0, \omega) \approx 1 + \frac{4\pi n e^2 \hbar^2}{2m E_g} \left(\frac{1}{E_g - \hbar\omega} + \frac{1}{E_g + \hbar\omega} \right), \tag{4.2.14}$$

and:

$$\epsilon(q = 0, \omega) \approx 1 + \frac{\hbar^2 \omega_p^2}{E_g^2 - \hbar^2 \omega^2} . \qquad (4.2.15)$$

At high frequencies ($E_g << \hbar\omega$), $\epsilon(q = 0, \omega)$ resembles the dielectric constant of a metal:

$$\epsilon(q = 0, \omega) \approx 1 - \frac{\omega_p^2}{\omega^2} . \qquad (4.2.16)$$

In this limit, metals and insulators have similar optical properties. The bulk plasmon energy $\hbar\omega_b$ is slightly larger than $\hbar\omega_p$:

$$\hbar\omega_b = \sqrt{\hbar^2 \omega_p^2 + E_g^2} , \qquad (4.2.17)$$

i.e.:

$$\hbar\omega_b = \hbar\omega_p \sqrt{\frac{\epsilon_\infty}{\epsilon_\infty - 1}} . \qquad (4.2.18)$$

The difference between $\hbar\omega_b$ and $\hbar\omega_p$ is generally weak. In MgO, $\hbar\omega_p = 21$ eV and the experimental bulk plasmon energy is 22.6 eV. Equation (4.2.18), which yields $\hbar\omega_b = 25.7$ eV in this oxide, most of the time over-estimates $\hbar\omega_b$.

The plasmon dispersion is obtained from the roots of $\epsilon(q, \omega) = 0$. Cappellini *et al.* (1993) have generalized their expression (4.2.9) for the static dielectric constant of semi-conductors to include dynamic processes, in a one-pole approximation (Hybertsen and Louie, 1986):

$$\epsilon(q, \omega) = 1 + \frac{1}{1/(\epsilon_\infty - 1) + \zeta(q/q_{TF})^2 + \hbar^2 q^4 / 4m^2 \omega_p^2 - \omega^2 / \omega_p^2} . \qquad (4.2.19)$$

For small q values, they have obtained a quadratic dependence of the plasmon frequency ω_b upon q, as in metals:

$$\hbar\omega_b = \hbar\omega_p \sqrt{\frac{\epsilon_\infty}{\epsilon_\infty - 1}} + \zeta_D \frac{\hbar q^2}{m} . \qquad (4.2.20)$$

The ζ_D constant is equal to 0.38 in Si and 0.34 in GaAs.

Local field effects

In screening processes, local field effects are the manifestation of the spatial inhomogeneities of the material and, in particular, its discrete atomic structure. The response to a charge localized in $\vec{r}_0$ depends upon both $\vec{r}$ and $\vec{r}_0$ and not only upon $\vec{r} - \vec{r}_0$.

Hybertsen and Louie (1987a; 1987b) have calculated the charge density induced by an applied external field and a test charge located at various positions in the unit cell, in silicon, germanium and LiCl crystals, using the density functional theory. In an effective one-electron approximation, they have estimated the electronic susceptibility matrix, which is a generalization of (4.2.3) to inhomogeneous sytems:

$$\chi_{\vec{G}\vec{G}'}(\vec{q}) = -\sum_{\vec{k},\lambda,\lambda'} \frac{f_0(\vec{k},\lambda) - f_0(\vec{k}+\vec{q},\lambda')}{E(\vec{k}+\vec{q},\lambda') - E(\vec{k},\lambda)}$$

$$\times \langle \vec{k},\lambda| \exp(-i(\vec{q}+\vec{G})\vec{r})|\vec{k}+\vec{q},\lambda'\rangle \langle \vec{k}+\vec{q},\lambda'| \exp(i(\vec{q}+\vec{G}')\vec{r})|\vec{k},\lambda\rangle .$$

$$(4.2.21)$$

They have deduced the value of the inverse dielectric matrix, $\epsilon^{-1}_{\vec{G}\vec{G}'}(\vec{q})$, including direct and exchange electron–electron interactions. In Si and Ge, they have shown that the dipole induced by an external field is not parallel to this field, because of the directionality of the sp3 hybrid orbitals, and that the polarizability along a bond is larger than in the perpendicular direction. In LiCl, which is an ionic insulator, they have found that, in the presence of a test charge, the induced charge is larger on chlorine atoms than on lithiums, due to the different values of the atomic polarizabilities of the species. The authors conclude that in semi-conductors and insulators, local field effects are important both qualitatively and quantitatively, because they greatly modify the screening characteristics.

While a good description of local field effects is difficult to achieve with the methods associated with the density functional theory, which work in Fourier space, it is straightforwardly obtained when the wave functions are developed on a localized orbital basis set, as in tight-binding approaches. To illustrate how local field effects show up in a self-consistent tight-binding approach, let us first consider the expression of the dielectric constant matrix on an atomic orbital basis set. In a self-consistent approach, each renormalized atomic energy level ϵ_i is a linear function of the charges $-(N_j - Z_j)$ borne by the atoms. For example, in a Hartree approximation, ϵ_i is equal to:

$$\epsilon_i = \epsilon_i^0 + U_i(N_i - Z_i) + \sum_j V_{ij}(N_j - Z_j) , \qquad (4.2.22)$$

with U_i the intra-atomic electron–electron repulsion integral and V_{ij} the inter-atomic interaction at a distance R_{ij}. A modification δN_j in the outer electron numbers induces changes $\delta \epsilon_i$ in the atomic energies:

$$\delta \epsilon_i = \sum_j \gamma_{ij} \delta N_j , \qquad (4.2.23)$$

with $\gamma_{ij} = \delta_{ij} U_i + V_{ij}$ (δ_{ij} is the Kronecker symbol). On the other hand, the electron numbers N_i are obtained by solving the Schrödinger equation. In the linear response theory, i.e. assuming that the changes in the effective atomic energies ϵ_i are small, one may write:

$$\delta N_i = \sum_j \chi_{ij} \delta \epsilon_j \,. \tag{4.2.24}$$

χ_{ij} is the matrix form of the electronic susceptibility, expanded on an atomic orbital basis set (Priester *et al.*, 1988). Equations (4.2.23) and (4.2.24) may be written in a matrix form and gathered together to give:

$$\delta N = (I - \gamma\chi)^{-1}\delta N_{\text{ext}}$$
$$\delta\epsilon = (I - \gamma\chi)^{-1}\gamma\delta N_{\text{ext}} \,, \tag{4.2.25}$$

in response to an external charge perturbation (δN_{ext} a column vector), or:

$$\delta N = (I - \gamma\chi)^{-1}\chi\delta V_{\text{ext}}$$
$$\delta\epsilon = (I - \gamma\chi)^{-1}\delta V_{\text{ext}} \,, \tag{4.2.26}$$

in response to an external potential perturbation. The $(I - \gamma\chi)$ matrix is the tight-binding form of the dielectric constant operator.

Tight-binding approaches are thus able to treat screening effects provided that they used at a self-consistent level. The screening strength depends not only on the electronic susceptibility χ, but also on the competition between intra-atomic and inter-atomic interactions involved in γ. This is consistent with the discussion of the ionic charge in insulators, which was given in Section 1.4. From (4.2.25) and (4.2.26), it can be seen that the response to a test charge located on an anion is different from that induced when the test charge is located on a cation, even when these species occupy sites which are equivalent in the crystal. This comes from the different polarizabilities of the species, revealed by their different U values. This is the most obvious manifestation of local field effects.

In order to get deeper insight into the strength and spatial extension of screening effects in an insulator, one can develop an analytical approach in which the electronic susceptibility is estimated in first-order perturbation with respect to the resonance integrals β, and in which a single orbital per site is considered to describe the electronic structure (Harrison, 1980). Under these approximations, the anion and cation electron numbers then read:

$$N_{\text{C}} = \frac{2Z_{\text{C}}\beta^2}{(\epsilon_{\text{C}} - \epsilon_{\text{A}})^2} \,, \tag{4.2.27}$$

and:

$$N_{\text{A}} = N_{\text{A}}^{\text{max}} - \frac{2Z_{\text{A}}\beta^2}{(\epsilon_{\text{C}} - \epsilon_{\text{A}})^2} \,, \tag{4.2.28}$$

with Z_{C} and Z_{A} the cation and anion coordination numbers ($mZ_{\text{C}} = nZ_{\text{A}}$ in a compound of stoichiometry $C_m A_n$), and $N_{\text{A}}^{\text{max}}$ is the anion electron number associated with its maximal formal charge (e.g. six 2p electrons for an oxygen atom). $\epsilon_C - \epsilon_A$ is the self-consistent energy difference between the cation and anion effective atomic energies. Equations (4.2.27) and (4.2.28) may also be derived from (1.4.48) in the limit where β is much smaller than $\epsilon_C - \epsilon_A$. According to the definition of the electronic susceptibility χ, the diagonal elements of χ on the anion and cation orbitals are respectively equal to $\chi_{CC} = -2N_C/(\epsilon_C - \epsilon_A)$ and $\chi_{AA} = -2mN_C/n(\epsilon_C - \epsilon_A)$. The non-diagonal matrix elements are non-zero only when neighbouring sites are involved. They read: $\chi_{AC} = \chi_{CA} = 2N_C/Z_C(\epsilon_C - \epsilon_A)$. In this simplified approach, $\epsilon_C - \epsilon_A$ is the gap width Δ and the susceptibility matrix depends only on the gap, the electron transfer N_C and the oxide stoichiometry.

Applied to oxides, this model predicts that the charge induced by a point charge is strongly localized on the atom on which the test charge is placed, and that the charge modifications on the neighbouring atoms decrease very quickly with distance. For example, in rocksalts, in which the inter-atomic distance is denoted R, one finds that, on the site of the perturbation:

$$\delta N_0 = \left(\frac{\frac{2N_C}{\Delta}(U_A - \frac{1}{R})}{1 + \frac{2N_C}{\Delta}(U_A - \frac{1}{R})} \right) \delta N_{\text{ext}}, \qquad (4.2.29)$$

while in the first coordination shell:

$$\delta N_1 = \left(\frac{\frac{2N_C}{\Delta}(\frac{1}{2R} - U_A - \frac{4}{R\sqrt{2}})}{1 + \frac{2N_C}{\Delta}(U_A - \frac{1}{R})} \right) \delta N_{\text{ext}}. \qquad (4.2.30)$$

These results apply to the case where the test charge is located on an anion. When it is located on a cation U_A has to be replaced by U_C. With reasonable values for the inter-atomic distances and U parameters, it turns out that δN_1 is more than one order of magnitude smaller than δN_0. The screening length is thus very short and the screening may be considered as local. A first estimate of ϵ_∞ may be done by considering only the charge δN_0 induced on the perturbed atom:

$$\epsilon_\infty \approx 1 / \left(1 + \frac{\delta N_0}{\delta N_{\text{ext}}} \right). \qquad (4.2.31)$$

This quantity should not be compared directly to the macroscopic optical dielectric constant, although it should be of the same order of magnitude, because it characterizes *local* properties of screening in the material, and because it depends upon whether the perturbation is localized on an anion or on a cation. It will be denoted ϵ_{eff}^A and ϵ_{eff}^C in these two cases, respectively. Within the present model:

$$\epsilon_{\text{eff}}^A = 1 + 2\frac{mN_C}{n\Delta}(U_A - \frac{1}{R}). \qquad (4.2.32)$$

Similarly, when the test charge is located on a cation:

$$\epsilon_{\text{eff}}^{C} = 1 + 2\frac{N_C}{\Delta}\left(U_C - \frac{1}{R}\right). \tag{4.2.33}$$

A similar expression is found for the response to an external alternating potential applied on anions and cations, with $U_i - 1/R$ replaced by $U_C + U_A - 2\alpha/R$. This case is relevant to the analysis of the bulk ionic charges performed in Chapter 1, Equation (1.4.50). The screening is strong in compounds in which the anion–cation electron transfer N_C is strong (small gap, large resonance integrals) and in which the ratio $(U - 1/R)/\Delta$ is large. These trends are qualitatively followed in rocksalt oxides, rutile and quartz.

4.3 Surface dielectric constant

Close to a surface, the description of screening effects is more difficult than in the bulk, because the periodicity of the system is broken in one direction. Even for an electron gas, assumed to be homogeneous in a half-space, the dielectric function is non-local. It is characterized by two wave vectors $\vec{q}$ and $\vec{q}'$ with the same projection $\vec{q}_{\parallel}$ in the surface plane: $\epsilon(\vec{q}_{\parallel}, q_z, q_z', \omega)$. In the classical macroscopic limit, image effects and the value of the surface plasmon energy will be analyzed first. Then, the relationship between the surface electronic structure and the dielectric function will be discussed. Finally the spatial dependence of screening effects in the vicinity of a surface will be exemplified.

Classical macroscopic limit

In the most simplifying picture, a dielectric medium, characterized by its optical dielectric constant ϵ_{∞}, is assumed to be separated from vacuum by a planar barrier of infinite height (Fig. 4.2), and its electronic structure is assumed to be unchanged by the presence of the surface. One then neglects the penetration of the electrons into the vacuum which takes place on distances of the order of $1/k_F$, the existence of surface states, the surface gap-narrowing and the existence of a surface dipole. Under these approximations, Cinal *et al.* (1987) have calculated the potential induced by the presence of a point charge located close to a semi-conductor surface, using Penn's (1962) expression for the bulk dielectric constant. They obtained values for the total potential which are very close to those predicted by classical approaches everywhere in space, except in a thin layer, roughly 1 Å wide at the surface.

When a test charge Q is located in vacuum at $\vec{r}_0$ – the origin of coordinates O is chosen at the projection of the test charge position

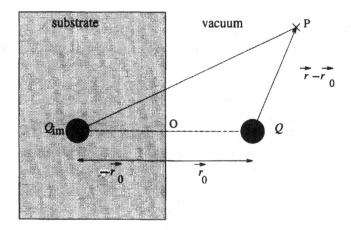

Fig. 4.2 External charge and image charge close to an interface.

on the interface – the electrostatic potential at $\vec{r}$ in vacuum has two contributions: the potential directly exerted by the charge Q, and a second term which comes from the reflection of the electric field on the interface. In the dielectric medium, the value of the potential is determined by the transmission of the electric field. The reflection and transmission coefficients are obtained by matching the tangential components of the electric field and the normal components of the electric induction on the interface. The result is the following for $\vec{r}$ in vacuum:

$$V(\vec{r}) = \frac{Q}{|\vec{r} - \vec{r}_0|} + \frac{(1 - \epsilon_\infty)}{(1 + \epsilon_\infty)} \frac{Q}{|\vec{r} + \vec{r}_0|} \,, \qquad (4.3.1)$$

and:

$$V(\vec{r}) = \frac{2}{(1 + \epsilon_\infty)} \frac{Q}{|\vec{r} - \vec{r}_0|} \,, \qquad (4.3.2)$$

for $\vec{r}$ in the dielectric medium. The reflected and transmitted potentials both involve the factor $1 + \epsilon_\infty$ in their denominator.

In vacuum, the potential may thus be calculated as if the polarizable medium was replaced by a charge located at $-\vec{r}_0$, called the image charge, whose value is:

$$Q_{\mathrm{im}} = \left(\frac{1 - \epsilon_\infty}{1 + \epsilon_\infty} \right) Q \,. \qquad (4.3.3)$$

The image charge represents the effect of dipoles which are induced by the external charge in the dielectric medium. It is fictitious. The dielectric medium remains electrically neutral.

The concept of image charge is widely used in the context of metal surfaces, when estimating work-functions, understanding image states or studying adsorption of charged species. In insulators or semi-conductors, it is less often invoked, because the image charges are considered to be smaller. However, they are far from being negligible. In insulators with the largest gaps ($\epsilon_\infty \approx 3$), they are roughly equal to half the external charge. It is thus necessary to take them into account in calculating adsorption energies for charged species or adhesion energies between two insulators.

As already mentioned, collective excitations are associated with a diverging dynamic response to a perturbation. Equations (4.3.1) and (4.3.2) suggest that a collective surface mode occurs when $1 + \epsilon(\omega) = 0$. By replacing $\epsilon(\omega)$ by its bulk expression (4.2.15), one obtains the surface plasmon energy $\hbar\omega_s$ equal to:

$$\hbar\omega_s = \sqrt{\frac{\hbar^2\omega_p^2}{2} + E_g^2} \, . \tag{4.3.4}$$

As a matter of comparison, we recall that the bulk plasmon energy is equal to:

$$\hbar\omega_b = \sqrt{\hbar^2\omega_p^2 + E_g^2} \, . \tag{4.3.5}$$

These expressions may also be used to estimate the surface and bulk plasmon energies in metals by setting $E_g = 0$. For MgO, in which $\hbar\omega_p = 21$ eV and $E_g = 14.8$ eV, one has $\hbar\omega_b = 25.7$ eV and $\hbar\omega_s = 20.9$ eV, values which are over-estimated with respect to the experimental values of 22.6 eV and 15.4 eV, respectively (Hengehold and Pedrotti, 1976). The agreement is worse for CdO and for ZnO. The reason for the disagreement may lie in the non-negligible values of the imaginary part of the dielectric constant and especially its too rapid variations for frequencies close to the plasmon pole.

A better account of the surface electronic structure

The above approach neglects the modifications of the electronic structure in the surface layers. In semi-conductors and insulators, as a result of the presence of dangling bonds or due to the reduction of the Madelung potential on under-coordinated atoms, surface states appear either at the gap edges or deep in the gap. Surface electron–hole pairs have thus lower energies than in the bulk. In addition, the reduction of the Madelung potential at the surface yields an increase of the matrix elements of γ in (4.2.23). Both processes enhance the surface dielectric constant.

Reining and Del Sole (1991a; 1991b) have analysed the shape of the screened potential on the Si(111) (2×1) reconstructed surface and the strength of dynamic screening effects. They have taken into account the electronic structure of the zig-zag silicon chains, using a one-dimensional tight-binding model, fitted to numerical calculations. They have found that the surface dielectric constant is about six times larger than the bulk value and that the screening is twice as strong, mainly because of the presence of surface states. They have predicted a value of the exciton binding energy equal to 0.3 eV, to be compared with a surface gap width equal to 0.75 eV, in their model.

The theoretical tight-binding model proposed in the previous section to stress the parameters on which the bulk dielectric constant relies, may be extended to surface effects. Still treating electron delocalization effects in first-order perturbation and neglecting the small induced charges on neighbouring atoms, the effective dielectric constant associated with the screening of a test charge located on a surface cation, for example, reads:

$$\epsilon_{\text{eff}}^{C} = 1 + 2\frac{N_C}{\Delta}(U_C - \frac{1}{R}) \left(\frac{\frac{Z_1}{\Delta_1^3} + \frac{Z_2}{\Delta_2^3}}{\frac{Z}{\Delta^3}} \right) . \qquad (4.3.6)$$

In (4.3.6), the surface cation is assumed to have Z_1 bonds in the surface plane and Z_2 bonds with anions located in the underlying plane ($Z_1 + Z_2$ is the surface coordination number, smaller than the bulk value Z) and Δ_1 and Δ_2 are the corresponding energy separations between cation and anion levels (Δ_1 and Δ_2 are smaller than the bulk energy separation Δ). Considering that surface charges are very close to the bulk charges ($Z_1/\Delta_1^2 + Z_2/\Delta_2^2 \approx Z/\Delta^2$), one may conclude that, due to the gap-narrowing, $\epsilon_{\text{eff}}^{C}$ increases as the surface gets more open. This is in qualitative agreement with the result obtained by Reining and Del Sole on Si(111) (2×1).

Screening processes also take place around adsorbed charged species. The adsorbate charge represents an electrostatic perturbation acting on the substrate atoms which has a more pronounced effect than a test charge when a true chemical bond forms. In Chapter 6, we will describe in detail the adsorption of protons and hydroxyl groups on oxide surfaces, and we will point out the parameters which drive the strength of the electron transfer between the adsorbate and the substrate. Here, we only wish to focus on the charge redistributions which result from this charge transfer. For example, when a chemical bond forms between a proton and MgO(001) (Goniakowski *et al.*, 1993), an electron transfer n_{H^+} from the surface oxygen to the 1s proton level takes place. If both atoms were isolated, n_{H^+} would be equal to the number of electrons δN lost by the surface oxygen. But $\delta N < n_{H^+}$ because screening effects try to wash out the important charge variation on the oxygen atom. The neighbouring magnesium atoms give $n_{H^+} - \delta N$ electrons, the largest part of which goes to the p_x and p_y oxygen orbitals. When the adsorption takes place on different oxides, it may be checked that the ratio $\delta N/n_{H^+}$ qualitatively varies as $1/\epsilon_\infty$. For example, in the series BaO(100), SrO(100), CaO(100), MgO(100), TiO$_2$(110) and SiO$_2$(0001), $\delta N/n_{H^+}=$ 0.61, 0.60, 0.55, 0.49,

0.01 and 0.45, while $1/\epsilon_\infty=$ 0.255, 0.305, 0.296, 0.332, $\approx$0.132, 0.417, respectively. The neighbouring magnesium charges, on the other hand, are not actually much modified – in agreement with the theoretical model depicted above – because they receive electrons from their own neighbours which are located even further from the adsorption site. Fig. 4.3 shows that this shift of electrons takes place because the energy differences between cation and anion effective levels (the local gaps) are alternatively larger and lower than in the clean substrate.

Let us denote by M_1 the cations which are located in the first coordination shell of the oxygen atom O_1 on which the proton adsorbs, by O_2 the oxygens located in the second coordination shell and so on. On O_1, a loss of electrons results from the O–H bond formation. It induces a strong lowering of the atomic levels (intra-atomic effects). On other substrate atoms, the proton electrostatic potential also shifts the levels towards lower energies, but the importance of the shift decreases with the distance. The energy separation between the O_1 and M_1 levels thus increases. The O_1–M_1 bond is more ionic and electrons are transferred from M_1 to O_1. On the M_1–O_2 bond, the level separation is smaller. The M_1–O_2 bond is more covalent and electrons are transferred from O_2 to M_1, and so on.

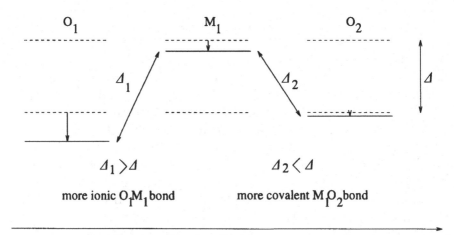

Fig. 4.3. Modification of the effective levels of the substrate atoms which are located close to an oxygen O_1, on which a proton is adsorbed. The atoms are denoted O_1, M_1, O_2, as their distance to the adsorption site increases. The local gap widths are enhanced and narrowed in an alternating way, which induces a shift of electrons towards O_1.

When a negatively charged species adsorbs on a surface, the electron motion has the opposite direction. Similar effects have been found in modelling tip–surface interactions (Castanier, 1995).

4.4 Conclusion

As in the case of metals and semi-conductors, there exist specific surface excitations in insulating oxides. Three types of surface phonon modes may be distinguished: the Rayleigh mode, the Fuchs and Kliewer modes and the microscopic surface modes. The first two modes have a long penetration length into the crystal. They are located below the bulk acoustic branches and in the optical modes, respectively. The latter are generally found in the gap of the bulk phonon spectrum.

Screening processes in insulators differ from those which occur in metals. The screening charge is not equal and opposite to the perturbing charge Q. Its value $Q(1/\epsilon_\infty - 1)$ is determined approximately by the value of the optical dielectric constant ϵ_∞, although one must remember than the screening is sensitive to the actual microscopic structure of the material, i.e. to local field effects. An expression for ϵ_∞ in terms of the anion–cation charge transfer N_C, the gap width Δ, and the relative value of the on-site and first-neighbour electron–electron interactions, qualitatively accounts for the screening properties of various oxides, both at the surface and in the bulk. The screening charge is mostly localized on the atomic site on which the test charge is located or on the substrate atom on which a charged particle adsorbs. Electrons can shift toward or from the perturbation, thanks to alternating variations of the energy separation between the anion and cation levels, which make the local bonds more or less ionic than in the absence of the perturbation.

The understanding of screening processes is of major importance in many circumstances in the physics of oxides. For example, in the bulk, it is known that defects yield gap states, which enhance the screening and favour dielectric breakdown. At the surface, annealing often induces a sub-stoichiometry in the outer layers, and strong electron redistributions take place around oxygen vacancies. The same is true in adsorption and adhesion processes.

5

Metal–oxide interfaces

5.1 Introduction

The research field of metal–oxide interfaces is very active, partly because of their important technological applications. For example, in hetero-geneous catalysis, oxide powders or porous compounds, such as zeolites, are used as supports for transition metal clusters, because they provide a large – external or internal – specific area of contact with the metal. In many cases, it is also recognized that they modify the cluster reactivity (Dufour and Perdereau, 1988). Oxide surfaces, such as those of MgO or SrTiO$_3$, whose quality and planarity are well controlled, have been used as substrates for the deposition of thin superconductor films. This has been particularly important since the discovery that some copper oxide based compounds remain superconductors above liquid nitrogen temperature. Thin metallic films are also deposited on various oxides in the fabrication of optical devices, or on glass in the fabrication of mirrors.

Oxides are often chosen as insulation materials, for example as sheaths for resistive heaters, due to their low electrical and thermal conductivity. In MOS transitors (MOS = Metal–Oxide–Semi-conductor), a thin SiO$_2$ layer is deposited between a doped silicon substrate and the metallic gate to control the channel conductivity. In more complicated electronic devices, with several integration levels, SiO$_2$ is also used to make insulating dielectric layers.

In order to confine very reactive liquid metals, with high fusion points, in containers or melting pots – for example, uranium in the nuclear industry – one uses oxides, which are present as natural layers on refractory metal surfaces, or other oxides such as Y$_2$O$_3$, which give the advantage of being thermodynamically stable, refractory and wear-resistant. The wetting of oxides by liquid metals is thus the object of much attention in this field, as well as in the fabrication of ceramic–ceramic or metal–ceramic joints with brazing alloys.

This quick overview stresses the importance of a detailed understanding, from a fundamental point of view, of the conditions under which a stable interface forms, the parameters which control the adhesion and more generally the atomic and electronic structure in the interfacial layers.

The deposition of clusters or thin metallic films on a surface may be performed in various ways. The vacuum condensation of a metallic vapour allows a precise control of the atom flux, of the substrate temperature and of its properties – its orientation, its defects, its homogeneity, its thickness, etc., while this is rarely the case when a chemical vapour deposition is performed. In order to make an interface between a metal and its own oxide, it is possible to reduce the oxide by atomic hydrogen, which often leads to good epitaxy and interface orientation. Conversely, a metallic surface may be oxidized by heating under an oxidizing atmosphere. In this case, the oxide is often amorphous. The kinetics of the oxide layer formation depend upon the diffusion of the metal and/or oxygen atoms through the oxide, and also upon the reaction rate at the interfaces (Mott and Gurney, 1948). Internal metal–oxide interfaces may be formed around precipitated oxide particles inside an alloy, as a result of oxidation of one or more of its constituents (Schmalzried and Backhaus-Ricoult, 1993). Finally, thermo-compression may also be used for solid–solid bonding (Akselsen, 1992a; 1992b).

The first stages of deposition from a vapour, are generally divided into three steps: adsorption–desorption, diffusion–migration, and nucleation and growth, schematized on Fig. 5.1.

Adsorption and desorption   In order to obtain a permanent condensation of metal atoms on a surface, the adsorbed flux must be larger than the desorbed flux. The characteristic time τ, which controls this process, is a function of the desorption free energy of activation, ΔG_{des}. The adsorption step is very sensitive to surface geometric or stoichiometric irregularities. As indicated in Fig. 5.1, decoration occurs when adsorption takes place preferentially on surface defects.

Migration and diffusion   The formation of small clusters requires a noticeable diffusion of the adsorbed atoms on the surface and efficient collisions, which are strongly dependent upon the substrate temperature and the presence of defects. At small sizes, the clusters have high probabilities of dissociation.

Nucleation and growth   Beyond a critical size of the clusters, growth is controlled by energetic parameters, which will be considered later in this chapter. Many different modes are observed, among which are:

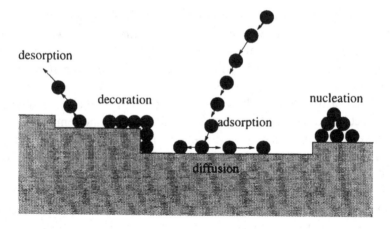

Fig. 5.1   Deposition of atoms on a surface, by condensation of a metallic vapour.

- The Frank–van der Merwe two-dimensional growth, in which layers are formed one after the other. This mode is sought to obtain thin deposits with a controlled thickness. Under some circumstances, a new layer may start forming before the preceding one is achieved. The growth is then said to follow a monolayer–simultaneous–multilayer mode.
- The island Volmer–Weber growth: clusters of all sizes are formed and a percolation occurs beyond a critical thickness.
- The Stranski–Krastanov mode: the first stage is a two-dimensional growth which leads to a single metal layer. Then cluster formation takes place.

These three modes are illustrated schematically on Fig. 5.2, upon which a fourth mode is also displayed, relevant for reactive interfaces. In the latter, an actual interphase is formed between the oxide and the metal as a consequence of a chemical reaction or inter-diffusion at the interface. The chemical composition of the interphase is generally intermediate between that of the metal and the oxide.

Growth modes have been studied by photoemission, Auger, electron energy loss and low-energy ion scattering spectroscopies on a number of metal–oxide interfaces. Through surface-sensitive techniques, one tries to determine whether the substrate surface is or is not completely covered, after a mean deposition thickness equivalent to one monolayer, and whether or not the signal presents a break in its slope for this coverage. It has been agreed for many years that Auger spectroscopy was one of the best tools in this regard (Argile and Rhead, 1989), but it is now recognized that, alone, it cannot be considered as a conclusive

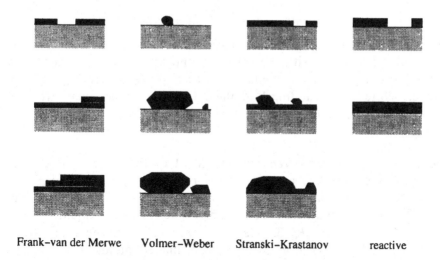

Frank–van der Merwe   Volmer–Weber   Stranski–Krastanov   reactive

Fig. 5.2   Four possible growth modes for a metal deposited on an oxide substrate.

determination of the growth mechanism (Diebold *et al.*, 1993b). The three-dimensional Volmer–Weber and the Stranski–Krastanov modes are often observed during the condensation of a metallic vapour on an oxide substrate, but at high temperature, when a liquid metal wets an oxide, the reactive mode may also occur. For some metal–oxide interfaces, no agreement has been reached in the literature upon whether the growth is of the Volmer–Weber or Stranski–Krastanov type. For example, this is the case of the Ag–MgO(100) interface (Palmberg *et al.*, 1967; Trampert *et al.*, 1992; Didier and Jupille, 1994a) and of the Cu–Al$_2$O$_3$ interface (di Castro *et al.*, 1985; di Castro and Polzonetti, 1987; Gautier and Duraud, 1994; Gota *et al.*, 1995; Møller and Guo, 1991; Chen *et al.*, 1992; Varma *et al.*, 1992). Such disagreements have their origin in the difficulties of interpretation of experimental results and in the comparison of systems prepared under different conditions and at different temperatures.

5.2 First stages of deposition

In the first stages of deposition, the metal atoms attach themselves randomly to the surface, before migrating, gathering and forming clusters or two-dimensional layers. Let us now discuss how these processes are affected by the geometric characteristics of the interface, the metal electronic parameters and the oxide stoichiometry, and consider a few examples of interfacial compound formation.

Atomic structure at the interface

The atomic structure at the interface depends upon the shape of the metallic clusters in contact with the substrate and upon the phase coherence between the deposit and the substrate lattice. As far as clusters are concerned, before understanding how the substrate modifies their structure, it is worthwhile recalling some of their atomic characteristics in the free state.

Geometric structure of free clusters Most of the properties of small clusters in the free state are determined by the outer atoms which have an incomplete coordination shell. Defining an elementary atomic volume $4\pi R_0^3/3$, and assuming that an n-atom cluster is like a liquid drop of volume $4\pi R^3/3 = n4\pi R_0^3/3$, one may estimate the number of surface atoms n_s, and the fraction of under-coordinated atoms in the cluster $f = n_s/n$. Here, n_s is equal to $4\pi R^2/\pi R_0^2$ and typical values of f are 0.86 and 0.4 for clusters of 100 and 1000 atoms. Up to a few hundred atoms, the cluster properties are thus mainly determined by their surface. However, the size criterion is only qualitatively valid, since the actual structure and shape of the particles are also important. Much work has been performed during the last ten years to understand the peculiarities of these small clusters (Joyes, 1990; Jortner, 1992):

- The atomic structure of a small cluster is not submitted to the same symmetry constraints as in bulk crystals. It may, for example, show local five-fold symmetries, which are forbidden in infinite periodic systems. For a given number of atoms, there often exists a large number of isomers, close in energy, which may change into each other as a result of thermal fluctuations.

- In metallic clusters, inter-atomic distances are shortened in the neighbourhood of the most under-coordinated atoms. In a macroscopic model of a liquid droplet, this contraction of the lattice parameter a may be accounted for by the effect of the Laplace pressure exerted by surface forces. It is a function of the particle radius R, of the compressibility coefficient K, of the surface tension γ and of the surface area A:

$$\frac{\Delta a}{a} = -\left(\gamma + A\frac{\mathrm{d}\gamma}{\mathrm{d}A}\right)\frac{2K}{3R}. \tag{5.2.1}$$

  At a more microscopic level, the arguments developed in Chapter 2 for surface relaxation, which are based on the competition between covalent energy and short-range atom–atom repulsion, also lead to contractions of bond-lengths in the neighbourhood of under-coordinated atoms.

- The electronic properties of small clusters, such as the density of states, the band width, the ionization potential, the gap width for insulators and semi-conductors, the plasma frequency and the elementary excitation spectrum, are also functions of the particle size. They vary in a non-monotonic way, because of shape effects, until the cluster reaches a critical size, and then converge monotonically towards the bulk properties.

Shape of supported clusters When clusters grow or are deposited on surfaces some of the peculiarities of isolated clusters persist, at least qualitatively, and they have to be considered in the interpretation of experiments. However, their shape is no longer uniquely determined by their surface energy σ, but also by the substrate surface energy σ_S and the interfacial energy σ_i. Winterbottom (1967) has written down the equilibrium condition for the shape of a particle on a substrate, at given volume, temperature and chemical potential:

$$0 = \delta \int_P \sigma^* dA . \qquad (5.2.2)$$

He has assumed that the interface is incoherent and that the corner and edge energies can be neglected at large enough sizes. In (5.2.2), the integration runs over the total surface of the particle and the effective surface energy σ^* is equal to σ at the vacuum interface and $\sigma_i - \sigma_S$ at the contact interface with the substrate. Determining the shape of a deposited particle is thus equivalent to finding the shape of a free particle having an anisotropic effective surface energy σ^*. The Wulf construction gives a graphic solution to such a problem. A 'σ-plot' is drawn in which the perpendicular distance h_{hkl} between the (hkl) surface and the particle centre is proportional to $\sigma^*_{(hkl)}$. Fig. 5.3 gives an example of a 'σ-plot' for a two-dimensional free particle of cubic symmetry.

The generalization of the Wulf construction to solid–solid interfaces leads to the discrimination of four cases (Fig. 5.4, *left column*), which correspond, respectively, to:

- non-wetting $\sigma_i - \sigma_S = \sigma$,
- partial wetting $0 < \sigma_i - \sigma_S < \sigma$,
- partial wetting $-\sigma < \sigma_i - \sigma_S < 0$. The effective surface energy σ^* is negative in this limit,
- total wetting $\sigma_i - \sigma_S < -\sigma$.

On the other hand, the knowledge of surface energies and interfacial energies is not sufficient to understand the growth dynamics at finite temperature. In their study of copper growth on the O-terminated polar ZnO(0001) face, Ernst et al. (1993) have, for example, stressed the importance of the evaporation energy of copper atoms from kinks located at

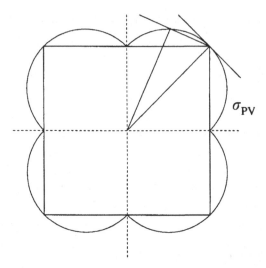

Fig. 5.3. Wulf construction for a two-dimensional free particle with cubic symmetry.

the cluster edge onto terraces. They have shown that, when this energy is larger than the difference between Cu–Cu and Cu–ZnO interactions, there may be efficient wetting at low coverage and low temperature, and three-dimensional clustering at higher temperatures.

Lattice coherence between the deposit and the substrate   Low-energy electron diffraction, high-resolution electron microscopy, surface EXAFS and SEELFS give structural information on the metal–oxide interface, such as the location of the metal atoms, the phase relationship between the particle and the substrate lattice, and the possible presence of dislocations which allow an accommodation of the elastic stress.

Palladium particles of about 20 Å diameter, deposited on single crystal MgO(100) surfaces, in ultra-high vacuum conditions, have been thoroughly studied in this respect. The comparison of the oxygen and magnesium K absorption edges, in the presence and in the absence of metal deposition, recorded by SEELFS (Goyhenex and Henry, 1992), reveals that the palladium atoms preferentially locate themselves on top of the substrate magnesium atoms. Under the conditions of the experiment, the palladium particles grow epitaxially, with a lattice parameter larger than that in the bulk, especially at small sizes (Giorgio *et al.*, 1990; 1993; Henry *et al.*, 1991). This result is at variance with the known contraction of inter-atomic distances in free metallic clusters. The maximal expansion of 8%, which is equal to the lattice misfit between bulk palladium and MgO, suggests that the pseudo-morphism between the two materials is

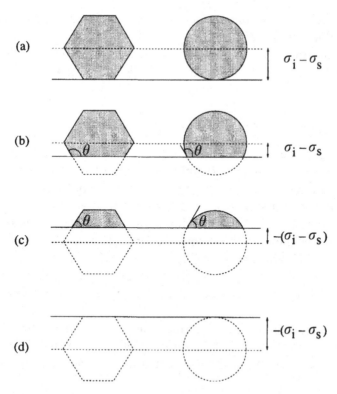

Fig. 5.4. Shape of small metallic particles (*left column*) or droplets (*right column*) on oxide surfaces for various wetting conditions: a) non-wetting; b) and c) partial wetting; d) total wetting.

responsible for the effects. Few systems involving metal clusters deposited on oxides have been the subject of such detailed studies.

On the contrary, many analyses have been performed on constituted interfaces, particularly those obtained by thermo-compression or internal oxidation. This is the case, for example, for metal–alumina interfaces, such as Al–Al_2O_3 (Timsit *et al.*, 1985), Ti–Al_2O_3 (Selverian *et al.*, 1991), Ni–Al_2O_3 (Trumble and Rühle, 1991), Fe–Al_2O_3 (Poppa *et al.*, 1986; Sakata *et al.*, 1986; Moon and Phillips, 1991; Epicier *et al.*, 1993), Nb–Al_2O_3 (Mader and Rühle, 1989), Pt–Al_2O_3 (Mudler and Klomp, 1985), Pd–Al_2O_3 (Gillet *et al.*, 1987; Eastman and Rühle, 1989), Cu–Al_2O_3 (Mudler and Klomp, 1985; Beraud and Esnouf, 1991), or for metal–MgO interfaces, such as Al–MgO (Höss *et al.*, 1990), Mo–MgO (Wu *et al.*, 1991b), Ni–MgO (Gao *et al.*, 1988) and Ag–MgO (Mader and Maier, 1990; Trampert *et al.*, 1992). In most cases, as revealed by high-resolution electron microscopy, the dense planes of both species align. When the misfit parameter $\eta = 2(a_1 - a_2)/(a_1 + a_2)$ is small, there exists an epitaxial relationship between the metal and the oxide and a dislocation network is present, which helps accommodate the elastic stress. The exact relative position of the metal and substrate atoms is rarely known. When η is large, the interface is often incoherent, especially when the interfacial energy is low.

Metal electronic parameters

XPS and Auger spectroscopies give information on the valence state of the metal atoms and on the possible reduction of the substrate cations. In a large number of systems, it has been found that interfacial bonds are formed in the first stages of deposition, leading to an oxidation of the deposited atoms, and that the metallic nature of the deposit is recovered only at larger thicknesses. This is the case for V–TiO$_2$ (Zhang and Henrich, 1992), Cu–MgO (Conard et al., 1992), Cu–SiO$_2$ (Xu et al., 1993), Fe–MnO (di Castro et al., 1993), Cu–ZnO (Møller and Nerlov, 1994), Ni–Al$_2$O$_3$ (Ealet et al., 1994), and Na–TiO$_2$ (Murray et al., 1995; Vermeersch et al., 1995).

Although each metal–oxide interface has its own characteristics, it is possible to derive some systematics for the growth mode and the interfacial properties, by considering a series of neighbouring metals in the periodic table.

Chromium, iron, copper and platinum deposits on TiO$_2$(110), obtained under ultra-high vacuum conditions, have been analyzed by LEIS (LEIS = low-energy ion scattering) and x-ray photoemission (XPS) (Diebold et al., 1993a; 1993b; 1994; Pan and Madey, 1993; Pan et al., 1993; Steinruck et al., 1995). LEIS results indicate that Volmer–Weber growth occurs for all metals. For a constant mean metal thickness, the percentage of rutile surface left free increases from chromium to copper. For copper, for example, at 30 Å thickness, about 37% of the rutile surface remains uncovered. In this series, the wetting is thus poorer and poorer. XPS experiments reveal two oxidation states of the titanium cations: Ti^{4+} and Ti^{3+}. The first corresponds to the titanium formal charge in rutile. The simultaneous presence of Ti^{3+} ions proves that a reduction of rutile by the metallic layer takes place. When the metal–oxide bond forms, the metal atoms are oxidized and transfer electrons to the substrate cations. In the series Cr, Fe and Cu, the XPS peak intensity associated with Ti^{3+} ions decreases, which suggests that the metal–oxygen bond weakens. Using macroscopic models which rely on surface energy considerations, these results may be rationalized, by assigning the changes in wetting behaviour to the changes in the interfacial energy σ_i, because the metal surface energy σ keeps a roughly constant value in the series. The strength of the metal–oxygen bond at the interface may be deduced from the formation energy of the metal oxide: the lower this energy, the lower the surface covering for a given deposition thickness. It strongly decreases in the series (Overbury *et al.*, 1975). The oxide formation energy and the interfacial charge transfer are functions of the difference in electronegativity between the metal and the oxygen. In the series this difference is equal to: $\chi_O - \chi_{Cr} = 1.9$, $\chi_O - \chi_{Fe} = 1.8$ and $\chi_O - \chi_{Cu} = 1.6$.

Oxide stoichiometry

Another factor which drives the growth characteristics of metallic deposits is the oxide surface stoichiometry.

On the (0001) face of α-alumina, for example, which shows various reconstructions as a function of the oxygen deficiency (Section 2.5), the observed copper growth mode remains three-dimensional (Volmer–Weber), whatever the surface stoichiometry, but the size of the copper clusters gets larger at higher reductions. The interpretation of these results relies upon the assumption that the surface oxygens are the centres for nucleation. The presence of oxygen vacancies induces a decrease of the cluster number and thus, for a given deposition thickness, an increase of their size (Gautier *et al.*, 1991a) .

For Cr, Ni, Cu and Pd depositions on α-alumina (Ealet, 1993), the strength of the metal–substrate interactions was found to be a function of the surface reduction: the higher the surface reduction, the stronger the Al–Pd and Al–Cu interactions, and the weaker the Al–Cr interaction. This suggests that for metal–substrate interfaces such that $\chi_M - \chi_{Al}$ is large and $\chi_O - \chi_M$ is low, the interactions are more sensitive to the surface reduction. The understanding of these effects is lacking.

The growth of iron films on TiO_2 surfaces submitted to an argon bombardment, was shown to be very sensitive to the surface roughness and the oxygen deficiency (Pan and Madey, 1993). The surface roughness prevents metal atom diffusion, and induces an improvement in the film spreading. On the contrary, oxygen sub-stoichiometry reduces the number of interfacial bonds and favours aggregation phenomena.

Examples of interfacial compound formation

In the examples quoted above, the formation of interfacial metal–oxygen or metal–cation bonds represents a preliminary step in the formation of an interfacial compound. However, more specific conditions – oxygen partial pressure, substrate temperature, etc. – are required in order that an actual interphase appears. For example, at the titanium–alumina($1\bar{1}02$) interface, the titanium is able to reduce efficiently the alumina surface at high substrate temperature and deposition thickness. A Ti_3Al interphase forms, which may be detected by Rutherford backscattering, electron microscopy and XPS (Selverian *et al.*, 1991). In a similar way, at 800 °C, under ultra-high vacuum and oxygen partial pressure, an $NiAl_2O_4$ spinel compound grows at the nickel–alumina interface (Zhong and Ohuchi, 1990).

5.3 Wetting and adhesion

Some important concepts for the understanding of interfacial forces have been developed in the field of high-temperature wetting by liquid metals.

General remarks

The wetting of a planar substrate by a liquid is characterized by the macroscopic contact angle θ (Fig. 5.5). The equilibrium condition for the contact line between the three phases: solid, liquid and vapour, yields the value of θ:

$$\cos \theta = \frac{\gamma_S - \gamma_i}{\gamma} . \qquad (5.3.1)$$

This Young's equation relates the wetting angle to the three surface tensions: the free metal surface tension γ, the free substrate surface tension γ_S and the interfacial tension γ_i.

The thermodynamic work of adhesion, on the other hand, is the work required to pull apart a unit area of a liquid–solid interface, thus creating one solid–vapour and one liquid–vapour interface:

$$W_{adh} = \sigma_S + \sigma - \sigma_i . \qquad (5.3.2)$$

Equations (5.3.1) and (5.3.2) yield the Young–Dupré relationship between the work of adhesion and the wetting angle:

$$W_{adh} = \sigma(1 + \cos \theta) , \qquad (5.3.3)$$

provided that the surface tensions γ and the surface energies σ are assumed to be equal. Depending upon the value of θ, four different cases, analogous to those met for solid particles, may be distinguished (Fig. 5.4, *right column*):

- no wetting $\theta = \pi$ when $\sigma_i - \sigma_S = \sigma$,
- bad wetting $\pi/2 < \theta < \pi$ when $0 < \sigma_i - \sigma_S < \sigma$,
- good wetting $0 < \theta < \pi/2$ when $-\sigma < \sigma_i - \sigma_S < 0$,
- total wetting $\theta = 0$ when $\sigma_i - \sigma_S < -\sigma$.

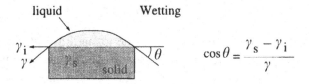

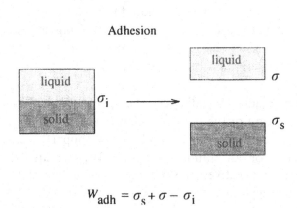

$$W_{\text{adh}} = \sigma_s + \sigma - \sigma_i$$

Fig. 5.5. Contact angle for a liquid droplet deposited on a solid planar surface and definition of the work of adhesion (the adhesion energy).

A measure of the contact angle thus gives an estimation of the work of adhesion, given the liquid surface energy. Equation (5.3.3) indicates that a strong adhesion force is associated with a small wetting angle θ, i.e. with a good wetting. For liquids with a low surface tension – typically less than 0.1 J/m^2 – dispersion forces or van der Waals forces are sufficient to realize a good wetting of the substrate by the liquid. But this is no longer true in the case of liquid metals, whose surface tension is one order of magnitude larger. The minimal interaction energy per interfacial atom required for total wetting ($W_{\text{adh}} = 2\sigma$) is then of the order of 1 eV, which corresponds to strong chemical bonds with covalent or ionic character.

Two families of metal–oxide interfaces are traditionally distinguished: the reactive and the non-reactive interfaces. At a non-reactive interface, the concentration of oxygen dissolved in the metal does not exceed the usual impurity concentration – a few parts per million (ppm). Chemical bonds between the oxide and the metal atoms may exist, but no displacement of atoms nor formation of a new interfacial compound takes place. If these latter effects occur, the interface is said to be reactive. The border line between the two families is ill defined, mainly because it is very temperature dependent.

Non-reactive interfaces

In the sessile drop method, a metal drop is deposited at high temperature onto a planar substrate, and its shape is used to determine the liquid metal surface tension σ and the contact angle θ. The experiments are performed under ultra-high vacuum or under an inert or a reducing atmosphere. The substrate cleanliness must be well controlled and the roughness must be less than 200 Å to ensure that the contact angle and the Dupré angle differ by less than 2°. Contact angles for liquid metals deposited on MgO, Al_2O_3 and SiO_2 are generally large and the corresponding adhesion energies low. Metal–oxide adhesion energies W_{adh} present some systematic trends as a function of the metal and oxide characteristics:

- On a given oxide, the adhesion energy of liquid metals belonging to the same column of the periodic table, decreases as the metal is located lower in the column. For example, for Group IB metals Cu, Ag and Au deposited on an alumina substrate, W_{adh} is equal to 490, 325 and 265 mJ/m^2, respectively. Similarly, W_{adh} is equal to 950, 345 and 245 mJ/m^2, respectively, for Al, Ga and In (Chatain *et al.*, 1986). The decrease of W_{adh} may be assigned to a steric effect, fewer metal atoms being present per interface unit area, as one goes down the series.
- For a given liquid metal deposited on various oxides, the adhesion energy increases as the oxide covalency grows. For example, Naidich (1981) quotes the following values for the contact angles of liquid gold on MgO, Al_2O_3, UO_2, Ti_2O_3, $TiO_{1.14}$ and $TiO_{0.86}$: 133°, 130°, 125°, 115°, 82° and 72°, a series along which the gap width decreases – Δ being equal to about 8 eV for the first two oxides, 5 eV for UO_2 and having lower and lower values for the titanium non-stoichiometric oxides. Similarly, on titanium oxides of various stoichiometries: TiO_2, Ti_4O_7 ($TiO_{1.75}$) and TiO, the contact angle of liquid gold decreases: 122°, 121° and 88° as the insulating character of the oxide weakens (Chabert, 1992; Chatain *et al.*, 1993).

These correlations make use of the atomic and electronic characteristics of the metal and of the oxide. However, the adhesion energy also depends upon the nature of the chemical bonds formed at the interface. This is the guiding idea of the first thermodynamical models of adhesion (McDonald and Eberhart, 1965), later improved by Chatain *et al.* (1986; 1987). The adhesion energy W_{adh} for an XO_n oxide, is related to the partial mixing enthalpies at infinite dilution of oxygens (O) and cations (X) in the metal

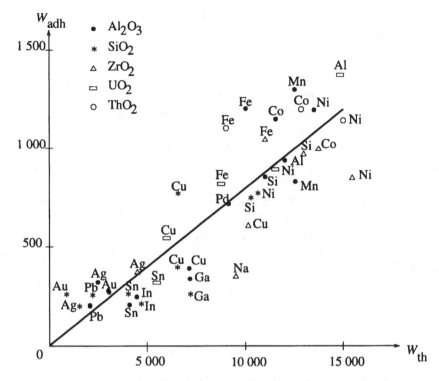

Fig. 5.6. Experimental values of the work of adhesion W_{adh} (in mJ/m^2) for pure metals on oxides, as a function of $W_{th} = (\Delta H^\infty_{O(M)} + \Delta H^\infty_{X(M)}/n)/V_M^{2/3}$ (in mJ/m^2) (according to Chatain *et al.*, 1986; 1987).

(M), $\Delta H^\infty_{O(M)}$ and $\Delta H^\infty_{X(M)}$:

$$W_{adh} = \frac{C}{V_M^{2/3}}\left(\Delta H^\infty_{O(M)} + \frac{1}{n}\Delta H^\infty_{X(M)}\right) . \tag{5.3.4}$$

V_M is the metal molar volume, which accounts for the steric effects at the interface, and C is a constant fitted to experimental values. This model takes care of the metal–oxygen and the metal–cation pair interactions through the quantities $\Delta H^\infty_{O(M)}$ and $\Delta H^\infty_{X(M)}$. Neglecting the entropy variations, a partial mixing enthalpy $\Delta H^\infty_{A(B)}$ may be written as a function of the AA, BB and AB pair interactions ϵ_{AA}, ϵ_{BB} and ϵ_{AB}, and as a function of the mean coordination number Z of an atom A dissolved in B, in the following way:

$$\Delta H^\infty_{A(B)} = Z\left(\epsilon_{AB} - \frac{\epsilon_{AA} + \epsilon_{BB}}{2}\right) . \tag{5.3.5}$$

For non-reactive systems, (5.3.4) is able to predict adhesion energies in reasonable agreement with experimental results (Fig. 5.6).

The wetting properties of oxides by liquid metals may be modified in various ways:

- By adding an element into the liquid metal, it is possible to decrease both σ_S and σ. The adsorption of the element at the solid–liquid interface generally improves both the adhesion and the wetting. On the contrary, its adsorption at the free liquid surface always destroys the adhesion. It improves the wetting only for θ angles lower than $90°$. These features have been checked on Al–Sn alloy–alumina interfaces (Eustathopoulos and Drevet, 1994).
- The adhesion characteristics drastically change when the oxygen partial pressure varies: new oxides may be formed, especially at low temperatures, or oxygen dissolution in the metallic phase may take place. In the latter case, one observes an improved wetting even for oxygen concentrations as low as a few ppm. This effect has been assigned to the precipitation of small oxide clusters at the interface and to the existence of a charge transfer between them and the substrate (Naidich, 1981).

Reactive interfaces

When atoms are displaced through the interface, the wetting is reactive. At metal–oxide contacts, the most common reaction is the reduction of the oxide by the metal, which leads, for example, to a new MO oxide:

$$M + XO \Rightarrow MO + X . \qquad (5.3.6)$$

The reaction takes place spontaneously if the total change in the Gibbs free energy ΔG_R^*, including the oxide formation Gibbs free energy ΔG_R^0 and the possible dissolution of the cations in the metallic matrix,

$$\Delta G_R^* = \Delta G_R^0 + \Delta H_{X(M)}^\infty , \qquad (5.3.7)$$

is negative. $\Delta H_{X(M)}^\infty$, in (5.3.7), is the partial mixing enthalpy at infinite dilution of cations X in the metal M. Fig. 5.7 shows the correlation between experimental wetting angles θ and calculated values of $\Delta G_R^*/RT$ for a series of metal–oxide interfaces. Qualitatively, an increase of $\Delta G_R^*/RT$ is associated with an increase of the wetting angle. The largest values for both quantities are found for non-reactive interfaces such as noble metals deposited on alumina or silica. Conversely, reactive wetting, with complete spreading of the liquid metal droplets, occurs for systems such as Ti and Zr on MgO.

Nevertheless, the correlation between $\Delta G_R^*/RT$ and θ is not perfect. Some interfaces, such as Sn–MgO and Cu–TiO, have identical $\Delta G_R^*/RT$

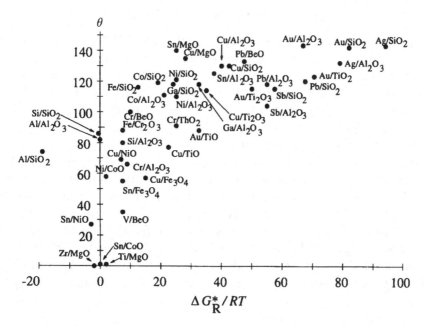

Fig. 5.7. Experimental values of contact angles θ (in degrees) for liquid metals on oxides, versus $\Delta G_R^*/RT$. Reactivity increases from the right to the left of the figure (according to Eustathopoulos and Drevet, 1994).

values and different contact angles ($\theta > 140°$ for Sn–MgO; $\theta = 80°$ for Cu–TiO). Similarly the contact angles of aluminium on silica and silicon on alumina are equal, while the $\Delta G_R^*/RT$ values are, respectively, negative and positive. To improve the analysis, Laurent (1988) and Eustathopoulos and Drevet (1994) have proposed an empirical relationship between the minimal wetting angle in a reactive system and the variation of free energy per unit area ΔG_R^* in the immediate neighbourhood of the interface:

$$\cos \theta_{\min} = \cos \theta_0 - \frac{\Delta \sigma_R}{\sigma} - \frac{\Delta G_R^*}{\sigma} . \qquad (5.3.8)$$

θ_0 is the contact angle in the absence of reaction and $\Delta \sigma_R$ is introduced to account for the influence of chemical effects and of the possible existence of interfacial reaction products, on the wetting.

5.4 Theoretical analysis of the first stages of deposition

Theoretical studies have been devoted to various aspects of the first stages of deposition, such as the electronic structure of supported clusters, the electronic structure of epitaxial monolayers and the modelling of growth modes.

Electronic structure of supported clusters

Using the DV-Xα method, Johnson and Pepper (1982) have calculated the electronic structure of one iron, nickel or copper atom adsorbed on a sapphire AlO_6^{9-} cluster. They find evidence of a hybridization between the metal d states and the substrate oxygen orbitals. In the series, the electrons fill all bonding states and an increasing number of anti-bonding states, which yields a weakening of the interfacial bond. Metal–cation bonds are not taken into account in this approach.

Anderson *et al.* (1987) have studied platinum–alumina interfaces for two terminations – aluminium and oxygen – of the alumina substrate. Two (111) layers of platinum were modelled, containing 15 or 16 atoms, and, depending upon the termination, the oxide is represented by an $Al_9O_{24}^{21-}$ or an $Al_{10}O_{18}^{6-}$ cluster. A strong interfacial bond, leading to a noticeable charge transfer, is evidenced between the platinium orbitals and the unsaturated aluminium orbitals. On the contrary, hardly any bonding and no charge transfer take place between the platinum and the oxygen orbitals.

Electronic structure of epitaxial monolayers

Several authors have studied the electronic structure of interfaces resulting from a two-dimensional epitaxial growth of metallic layers on an oxide surface.

On the (0001) face of α-alumina covered with one niobium, zirconium, molybdenum, ruthenium or palladium monolayer, Ohuchi and Koyama (1991) have shown that the metal d band is located in the energy range of the alumina gap. It is hybridized with the oxygen states of the valence band. In the series, more and more anti-bonding d states are filled, which yields results in agreement with the cluster calculations quoted above.

The characteristics of three layers of transition metals of the first series, epitaxially grown on the same Al_2O_3 surface, terminated by an oxygen layer, are very similar (Nath and Anderson, 1989). The metal–oxygen bond energy decreases in the series, while the interfacial charge transfer presents non-monotonic variations. It ranges from 0.45 to 0.72 electrons per surface oxygen. The adsorption of Cu on α-$Al_2O_3(10\bar{1}0)$ surfaces also displays a preferential bonding between the copper atoms and the surface oxygens (Kasowski *et al.*, 1988). For the Ni(111)–Al_2O_3 interface it was argued, on the other hand, that a strong hybridization between the aluminium surface state and the nickel orbitals takes place (Hong *et al.*, 1990).

The electronic and geometric characteristics of the Ag–MgO(100) interface have aroused much interest. Freeman *et al.* (1990), Blöchl *et al.* (1990) and Li *et al.* (1993) have found that the silver atoms are more stable above the surface oxygens and that the silver–oxygen bonding induces the presence of a small oxygen contribution to the density of states at the Fermi level, despite a weak charge transfer. Schönberger *et al.* (1992) confirm that silver and titanium atoms bind to surface oxygens. The titanium–oxygen bond is stronger with a partial covalent character, while the silver–oxygen bond shows an ionic character. At variance with these results, in the atomistic model developed by Duffy *et al.* (1992), the interface is found more stable when the silver atoms are located on-top the surface magnesiums or in bridging positions between two magnesiums. There is no consensus on this interface at present. One should note that the adsorption energies for the three adsorption sites differ by less than 0.04 eV, which is undoubtedly below the precision of the calculations, considering the assumptions made.

Growth mode modelling

No *ab initio* molecular dynamic method, such as the Car–Parrinello (1985) approach, has yet been applied to metal–oxide interfaces to determine the growth modes, because it would require too much memory and computation time. Several molecular dynamic approaches, making use of empirical potentials, have derived trends in the wetting, the growth and the nucleation modes of clusters on surfaces (e.g. Harding, 1992).

Among them, Grabow and Gilmer (1988) have obtained a phase diagram for the growth modes as a function of parameters which characterize the interface: the interfacial coupling W and the lattice misfit parameter η between the deposit and the substrate. For the inter-atomic interactions, they have used a 6–12 Lennard–Jones potential and a Stillinger–Weber potential. The latter, more suited to silicon, is a classical pair potential with a three-body angle-dependent term added, which favours the tetrahedral structure. Grabow and Gilmer find that the Volmer–Weber growth is restricted to systems in which the deposit–substrate interaction W is weak (Fig. 5.8). Beyond a critical value W_c, a layer-by-layer growth takes place. However, the lattice misfit η induces stresses in the film which destroy its thermodynamic stability as η increases. Beyond a given thickness, ranging from 2 to 5 monolayers, which is an increasing function of $1/\eta$, clusters are formed. This is a Stranski–Krastanov growth mode. The Frank–van der Merwe (layer-by-layer) mode is only seen for $W > 1$, under perfect expitaxy conditions: $\eta = 0$.

To conclude, one is very far from having a thorough picture of the first stages of formation of a metal–oxide interface. Except in very few cases, one ignores the nature of the adsorption sites and the number of interfacial bonds. For large size clusters, there is no clear criterion for

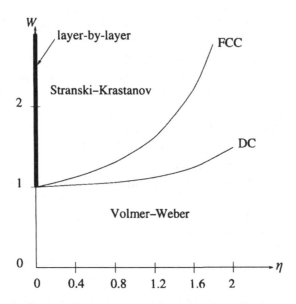

Fig. 5.8. Phase diagram for three growth modes: layer-by-layer, Volmer–Weber and Stranski–Krastanov, as a function of the interfacial coupling W and the lattice misfit parameter η between the deposit and the substrate. The phase limits denoted FCC (face centred cubic) and DC (diamond cubic) correspond to the use of Lennard–Jones and Stillinger–Weber potentials, respectively. The layer-by-layer growth mode is restricted to the axis $\eta = 0$ in the region $W > 1$ (after Grabow and Gilmer, 1988).

obtaining perfect epitaxy. The bond strength at the interface is related to the electronegativities of metal atoms and cations, and to the filling of the d band when transition metal atoms are involved, but one is not able to predict which effect prevails nor how the sign of the charge transfer is determined. Finally, the influence of surface defects, such as the surface roughness or the oxygen deficiency, on the film spreading is ill understood, despite its obvious technological interest.

5.5 Description of a jellium–oxide interface

We now present a theoretical model suited to non-reactive adhesion, which describes the electronic states at the interface between a thick metallic overlayer and an oxide, as a function of the metal and oxide characteristics (Bordier and Noguera, 1991; 1992).

The model

The oxide electronic structure is calculated via a tight-binding approach. The one-electron Hamiltonian is projected on an atomic orbital basis set, assumed orthonormal, which includes one non-degenerate s orbital per site. The orbital energies are denoted by ϵ_A and ϵ_C, respectively, for the anions and the cations. The electron resonant integrals β between neighbouring atoms are isotropic. The wave functions are indexed by the $k_\parallel$ component of the Bloch vector parallel to the surface. Two crystal structures are considered: the NaCl and the ZnS structures, and for each of them, the most stable surface is chosen – NaCl{100} and ZnS{110}. As proved in Section 1.4, the gap width is equal to $\epsilon_C - \epsilon_A$ under these hypotheses and the ratio $(\epsilon_C - \epsilon_A)/\beta$ fixes the degree of ionicity of the anion–cation bond in the substrate.

The metal, assumed to be a simple sp metal, is modelled by a jellium, with plane waves as eigenfunctions. All the metal electronic properties are expressed as a function of the electron density and as a function of a reference energy: the bottom of the conduction band E_{CB}, or the work-function Φ_M.

Metal-induced gap states (MIGS)

In order to simplify the wave-function matching, an abrupt planar interface between the metal and the oxide is assumed. In the gap energy range, the metal gives rise to damped waves which enter the oxide within a small region close to the interface (Fig. 5.9). These quantum states are called MIGS (MIGS = Metal-Induced Gap States). The existence of MIGS is not restricted to interfaces between simple metals and oxides. MIGS have also been found at the Ag–MgO interface (Blöchl *et al.*, 1990) and at metal–semi-conductor interfaces (Louie *et al.*, 1976; Tersoff, 1985). The variations of the MIGS density $N(E, z)$ in the oxide half-space, displayed on Fig. 5.9, depend upon the MIGS energy E. Close to the gap edges, the penetration length is large and when the energy reaches the top of the valence band or the bottom of the conduction band, the MIGS gradually transform into propagating band states. On the contrary, at the gap centre, the damping length l_p is of the order of a few inter-atomic distances. It decreases as $\epsilon_C - \epsilon_A$ gets larger, but varies little with the metal electron density. At the gap centre, the amplitude of the MIGS at the position of the interface $n_0 = N(E = (\epsilon_C + \epsilon_A)/2, z = 0)$ depends little upon the oxide characteristics. Fig. 5.10a and 5.10b show the variations of l_p and n_0 as a function of the oxide ionicity.

Charge transfer

In the oxide half-space, the MIGS formation can be correlated with a decrease in the number of valence and conduction band states. The oxide

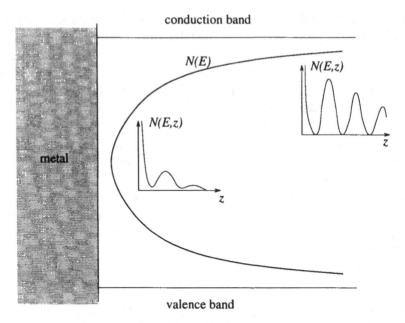

Fig. 5.9. MIGS density of states in the oxide gap. The inserts display the MIGS density in the oxide half-space for two energies: one near a gap edge and the other at the centre of the forbidden gap.

remains electrically neutral if the MIGS are filled up to some energy, called the zero-charge point E_{ZCP}. In the model developed here, E_{ZCP} is exactly located at mid-gap: $E_{ZCP} = (\epsilon_C + \epsilon_A)/2$.

For most metal–oxide interfaces, however, the Fermi level does not coincide with E_{ZCP}. A charge transfer takes place, which aligns the chemical potentials, and induces an interfacial dipole potential, which bends the bands. It is possible to estimate the self-consistent charge density in the vicinity of the interface, within a Thomas–Fermi approximation, if the MIGS density at mid-gap is taken equal to a single exponential function: $N(E_{ZCP}, z) = n_0 \exp(-z/l_p)$. The potential $V(z)$ due to the mean charge density $\rho(z)$ is related to $\rho(z)$ by Poisson's equation:

$$\frac{\mathrm{d}^2 V}{\mathrm{d}z^2} = -\frac{\rho(z)}{\epsilon} \,, \tag{5.5.1}$$

in which the dielectric constant ϵ is equal to ϵ_0 in the metal ($z < 0$) and $\epsilon_0 \epsilon_I$ in the oxide ($z > 0$). If $N_M(E_F)$ denotes the metal density of states at the Fermi level, the charge density reads:

$$\rho(z) \approx -eN_M(E_F)(-eV(z)) \,, \tag{5.5.2}$$

on the metal side, and:

$$\rho(z) \approx -en_0 \exp(-z/l_p)(E_F - E_{ZCP} + eV(z)) \,, \tag{5.5.3}$$

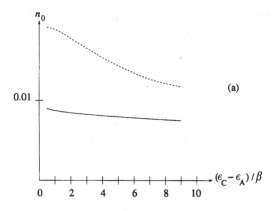

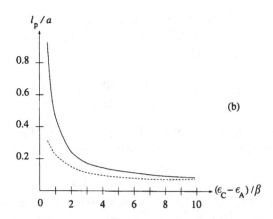

Fig. 5.10. MIGS density of states at the interface n_0 and MIGS damping length l_p, for energies at the gap centre, as a function of the oxide ionicity $(\epsilon_C - \epsilon_A)/\beta$, for two crystal structures (NaCl: full lines; ZnS: dashed lines). The anion level position ϵ_A is taken at 6 eV above the bottom of the metal conduction band and the oxide lattice parameter is $a = 5$ Å (after Bordier and Noguera, 1991).

on the oxide side. One should note that, in a Thomas–Fermi approximation, it is not necessary to know precisely the modifications of the density
of states in the valence band to calculate $\rho(z)$, because only energies close
to the Fermi level are relevant. The screening effects in the metal and
the insulator are treated on an equal footing, because, within a distance
l_p from the interface, the MIGS confer a quasi-metallic character to the
insulator. The latter thus presents two types of polarizability: an ionic
polarizability associated with the dielectric constant ϵ_I, and a metallic polarizability due to the presence of zero-energy excitations, whose strength
is driven by the MIGS density.

Due to the simple expressions of the local densities of states, the solution of Poisson's equation can be performed analytically. As a function of the modified Bessel functions I_n and of the Thomas–Fermi lengths l_M and l_I, respectively, for the metal and the insulator ($l_M = \sqrt{\epsilon_0/e^2 N_M(E_F)}$ and $l_I = \sqrt{\epsilon_0\epsilon_I/e^2 n_0}$), the electrostatic potential is given by:

$$eV(z) = (E_{ZCP} - E_F) + \alpha I_0 \left(\frac{2l_p}{l_I} \exp(-z/2l_p) \right) , \qquad (5.5.4)$$

if $z > 0$ (insulator side), and:

$$eV(z) = \alpha' \exp \left(\frac{z}{l_M} \right) , \qquad (5.5.5)$$

if $z < 0$ (metal side). The α and α' constants are determined by matching the electrostatic potential and the electric induction at $z = 0$. For example, on the oxide side of the interface:

$$eV(z) = (E_{ZCP} - E_F) \left(1 - \frac{I_0[2l_p \exp(-z/2l_p)/l_I]}{I_0(2l_p/l_I) + (l_M\epsilon_I/l_I)I_1(2l_p/l_I)} \right) . \qquad (5.5.6)$$

$V(z)$ is proportional to the electrostatic perturbation ($E_F - E_{ZCP}$), times a function which depends upon the MIGS density $n_0 l_p$. When $n_0 l_p$ is large, the ratio between the Thomas–Fermi lengths l_p/l_I is high and the interfacial potential tends towards $eV(+\infty) = E_{ZCP} - E_F$. The bands are bent in order to align the Fermi level with the zero-charge point, deep inside the insulator. This limit is reached for covalent oxides. In the opposite case of highly ionic oxides, $eV(+\infty) = 0$, and the Fermi level is fixed by the metal. There is no charge transfer nor band-bending at the interface. Fig. 5.11 shows the most general band diagram.

The Schottky barrier height

The Schottky barrier height Φ_B is the energy required to excite an electron from the metal Fermi level to the bottom of the insulator conduction band, far from the interface. The Schottky barrier height and the interface index $S = \partial\Phi_B/\partial\chi_M$, which represents its derivative with respect to the metal electronegativity χ_M, have been measured for a large number of metal–semi-conductor interfaces. It has been shown that, for a given semi-conductor, Φ_B is roughly a linear function of χ_M, so that S is uniquely determined by the characteristics of the semi-conductor. It increases as the ionicity of the latter grows.

Assuming that the band-bending and the interfacial charge transfer are negligible, Schottky (1939) wrote Φ_B as the difference between the metal

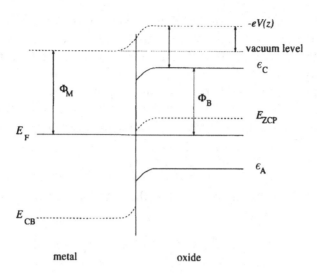

Fig. 5.11. Band-bending at a metal–oxide interface: the metal is on the left and the oxide on the right. ϵ_C and ϵ_A are the bottom of the conduction band and the top of the valence band, respectively.

work-function Φ_M and the insulator electron affinity χ (Fig. 5.11):

$$\Phi_B^{Sch} = \Phi_M - \chi . \qquad (5.5.7)$$

Subsequently, a number of experiments have proved that Φ_B is not given by Schottky's equation, and that it does not depend upon the metal properties. Such a result may be found if the Fermi level is pinned by a surface state band (Bardeen, 1947), in which case Φ_B is equal to the energy difference between the bottom of the conduction band and the surface state. The same is true in the presence of defect states or of high MIGS densities, such as those seen at the interfaces between metals and covalent semi-conductors (Rhoderick, 1978; Schlüter, 1982; Brillson, 1982; Flores and Tejedor, 1987).

The analytical approach developed above yields an expression of Φ_B as a function of the MIGS density, which is valid whatever the ionicity of the insulator:

$$\Phi_B = \epsilon_C - E_F - eV(+\infty) . \qquad (5.5.8)$$

When the MIGS density is high, $eV(+\infty) = E_{ZCP} - E_F$, according to (5.5.6), and:

$$\Phi_B = \Phi_B^{Bard} = \epsilon_C - E_{ZCP} . \qquad (5.5.9)$$

In this limit, Φ_B depends only upon the insulator characteristics, because the Fermi level is pinned at the zero-charge point E_{ZCP}. This is similar to Bardeen's result, although the nature of the states which pin the Fermi level E_F is different. When the insulator ionicity grows, the MIGS density decreases and becomes insufficient to screen the electrostatic perturbation $E_F - E_{ZCP}$. In the extreme ionic limit, $V(\infty)$ is equal to zero, in agreement with Schottky's model. It is crucial, in such a problem to describe adequately the position of the Fermi level in the gap. Indeed, in the literature, only self-consistent approaches have been able to yield reasonable values of the interface index S, but they were restricted to systems of low (Louie and Cohen, 1975; Louie et al., 1976; 1977) or very large (Feuchtwang et al., 1982; Pong and Paudyal, 1981) ionicity.

The most general expression of Φ_B, valid in the whole ionicity range, is obtained by combining (5.5.6) for the interfacial potential with the definition (5.5.8) of Φ_B:

$$\Phi_B = \frac{\Phi_B^{Sch}}{w} + \Phi_B^{Bard}\left(1 - \frac{1}{w}\right), \tag{5.5.10}$$

with:

$$w = I_0\left(\frac{2l_p}{l_I}\right) + \frac{l_M}{l_I}\epsilon_I I_1\left(\frac{2l_p}{l_I}\right). \tag{5.5.11}$$

It is a weighted average between Φ_B^{Sch} and Φ_B^{Bard} which refer, respectively, to the ionic ($w = 1$) and covalent ($w \to \infty$) limits, i.e. to the Schottky and Bardeen limits, (5.5.7) and (5.5.9). The interface index S is equal to:

$$S \approx \frac{\partial E_F/\partial \chi_M}{w}. \tag{5.5.12}$$

The variations of S as a function of $(\epsilon_C - \epsilon_A)/\beta$ given by (5.5.11) and (5.5.12) are in good qualitative agreement with experimental results, which represents an improvement upon previous approaches (Bordier and Noguera, 1991).

For strongly ionic insulators, a polynomial expansion of the Bessel functions yields the following expression for S:

$$S \approx \frac{\partial E_F/\partial \chi_M}{1 + (\epsilon_I l_p/l_I^2)(l_p/\epsilon_I + l_M)}. \tag{5.5.13}$$

Cowley and Sze's (1963) capacitor model is valid in this limit:

$$S \approx \frac{\partial E_F/\partial \chi_M}{1 + e^2 D_S l_{eff}/\epsilon_0}. \tag{5.5.14}$$

These authors have noted the analogy between an interface and a capacitor of thickness l_{eff}, whose plates bear a charge density D_S. Comparing the microscopic (5.5.13) and empirical (5.5.14) equations, one identifies l_{eff} with: $l_p/\epsilon_I + l_M$ and D_S with the MIGS density $n_0 l_p = \epsilon_I \epsilon_0 l_p/l_I^2$.

In the covalent limit, the MIGS penetration length l_p is no longer the relevant length scale. The charge penetration inside the insulator is limited by the screening length l_I, and the interface index is equal to:

$$S \approx \frac{\partial E_F}{\partial \chi_M} \sqrt{\frac{4\pi l_p}{l_I}} \frac{\exp(-2l_p/l_I)}{(1 + \epsilon_I l_M/l_I)}. \tag{5.5.15}$$

In this limit, the expression proposed by Cowley and Sze, although very often used, is not valid.

The quantities Φ_B and S depend upon the insulator ionicity through the ratio $(\epsilon_C - \epsilon_A)/\beta$, which fixes l_p and l_I. In the literature, the Schottky barrier height and the interface index have been compiled instead as a function of the anion–cation electronegativity difference (Kurtin *et al.*, 1969), but this was later criticized by Schlüter (1978) and Cohen (1979).

The interfacial potential $V(z)$ shifts the core levels of the atoms located close to the interface, as well as their outer orbitals. Yet, most analyses of core-level shifts neglect the band-bending effects and take the vacuum level as the reference energy (Quiu *et al.*, 1987). Such an assumption is justified only for highly ionic substrates.

The jellium–oxide interface model makes strong assumptions about the interface geometry and the band structures. It considers that metal–oxide contact is perfect and planar with no geometric or stoichiometric defect. However, actual interfaces present steps, roughening or oxygen vacancies, as a result of the surface preparation. Specific states, located in the insulator gap are associated with these features, and may take part in the charge-transfer process, as the MIGS do (Spicer *et al.*, 1980; 1986). In addition, in the oxide, the hypothesis of isotropic anion–cation resonant integrals $ss\sigma$, has strong implications on the surface states, because it prevents the existence of dangling bonds. While at the rocksalt oxide {001} surfaces no deep surface state exists, this is no longer true on the {110} faces of ZnS crystals or on the α-alumina(0001) face, terminated with aluminium atoms (Section 3.2). The existence of surface states is taken into account in the resonating dangling bond model developed for metal–semi-conductor interfaces (Mele and Joannopoulos, 1978; Lannoo and Friedel, 1991). Finally, the jellium approximation hides the concept of atom–atom bonding at the interface and makes a comparison with numerical or thermodynamical approaches difficult.

5.6 Interfacial energy

The estimation of the surface and interfacial energies σ, entering the expression for W_{adh}, is a key point in the understanding of wetting, adhesion and growth modes. Besides an entropic contribution, each σ contains electronic terms – kinetic, electrostatic, exchange and correlation,

Equation (1.4.64) – and a short-range repulsion term, some of which favour the adhesion and some oppose it. To the author's knowledge, no microscopic theory of metal–oxide adhesion has ever considered all these contributions. This is particularly true for the kinetic and electrostatic terms, induced by charge transfer at the interface, which have rarely been discussed in the past. A single quantum simulation of epitaxial monolayers on the α-alumina(0001) surface (Nath and Anderson, 1989) gives the values of the kinetic and short-range energies, but the electrostatic, exchange and correlation terms are absent. In this section, we wish to discuss (Noguera and Bordier, 1994) each of these terms by analogy with jellium–vacuum interfaces (Lang and Kohn, 1970).

Short-range repulsion energy

The exchange and correlation energies include a part of the short-range ion–ion repulsion due to the overlap of inner shells (Chapter 1). When only valence electrons enter the quantum treatment, this term has to be added afterwards. At surfaces, where a bond-breaking process takes place, it gives a negative contribution to the surface energy (Section 3.2). Its order of magnitude ranges from a few tenths of eV to several eV per broken bond.

Kinetic energy

The kinetic energy is associated with the quantum delocalization of electrons. It varies strongly with the geometry of the system. For example, when an electron is confined in a one-dimensional box of side L, $E \approx -1/L^2$. At a metal–vacuum interface, electrons spread into vacuum on a length scale which is inversely proportional to the Fermi wave vector. This lowers the total energy and thus brings a negative contribution to the surface energy – typically $-1.85 \ J/m^2$ for a magnesium–vacuum interface in the jellium model. This contribution may be compared to the total magnesium surface energy, which is equal to $0.6 \ J/m^2$ (Lang and Kohn, 1970).

The matching of the wave functions at a metal–insulator interface is similar to that which takes place on a free surface, although the insulator band structure has to be considered. In the gap energy range, the metallic waves give rise to metal-induced gap states and some valence and conduction band states disappear. The interfacial kinetic energy σ_{kin} is of the order of 10^{-1} to $1 \ J/m^2$, for typical MIGS densities of 10^{-2} states/(eV Å^3), a non-negligible value, when compared to the other terms. It is not easy to predict the sign of the kinetic contribution to the adhesion energy, because it depends upon the relative values of the interfacial and free-surface kinetic terms. It is nevertheless clear that a large interfacial kinetic energy (in absolute value) favours the adhesion.

Electrostatic energy

Two mechanisms are responsible for the electrostatic forces: the electron redistribution which induces the creation of an interfacial dipole and the polarization effects.

Interfacial dipole At a metal–vacuum interface, electron tunnelling into vacuum modifies the electron–electron interactions and weakens the electron–ion interaction (E_{ee} and E_{ec}, respectively, in Equation (1.4.64)). These two processes use up energy when a surface is formed. Lang and Kohn quote the following values : 0.43 J/m^2 and 0.5 J/m^2 at a magnesium–vacuum interface, although the latter, calculated in perturbation, is likely to be over-estimated

At a metal–oxide contact, there also exists an interfacial dipole, when the metal Fermi level does not coincide with the oxide zero-charge point. The electron–electron interaction term may be estimated using the analytical self-consistent approach presented in the previous section:

$$
\sigma_{ee} = \frac{1}{2} \int_0^\infty dz \rho(z) V(z)
$$
$$
= \frac{\epsilon_I \epsilon_0 (E_F - E_{ZCP})^2}{4e^2} \frac{2l_p(I_1^2 - I_0 I_2) + \epsilon_I l_M I_1^2}{(l_I I_0 + \epsilon_I l_M I_1)^2}.
\tag{5.6.1}
$$

In (5.6.1), the argument of the modified Bessel functions I_0, I_1 and I_2 is: $x = 2l_p/l_I$. The quantity σ_{ee} is always positive but may vanish if $E_F = E_{ZCP}$. The factor $(E_F - E_{ZCP})^2$ is a maximum for metals with a low work-function – alkaline metals – in contact with a wide-gap insulator, or for high work-function metals – noble metals, Au, Pt – deposited on a narrow-gap insulator. The second factor in the expression of σ_{ee} vanishes in the ionic limit, due to the low MIGS density and the absence of band bending. In the covalent limit, it reaches an asymptotic value equal to $\epsilon_0(E_F - E_{ZCP})^2/4e^2 l_{eff}$. It is the energy of a capacitor charged under a bias $V \propto (E_F - E_{ZCP})/e$, and whose plates are separated by a distance of $l_{eff} = l_M + l_I/\epsilon_I$, the sum of the screening lengths. An average value of 1 eV for $(E_F - E_{ZCP})$ corresponds to a typical electrostatic energy of the order of 10^{-2} to 10^{-1} J/m^2. To the author's knowledge this term has not previously been considered for metal–oxide interfaces.

The modifications of the electron–ion interactions in the presence of the interface also contribute to the electrostatic term. The total electrostatic energy gives the Madelung contribution to the interfacial energy. It is unfavourable to adhesion when it is high.

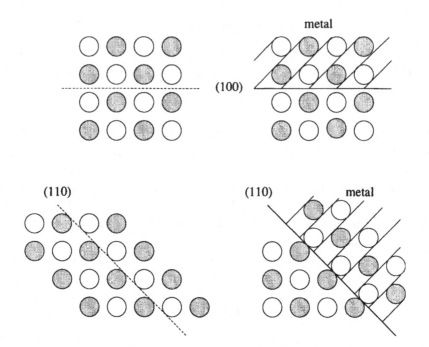

Fig. 5.12. Cut of a rocksalt oxide along the two orientations: {100} (*upper*) and {110} (*lower*). The left side of the figure represents a free surface. On the right side are represented the electrostatic images of the oxide ions at a metal–oxide interface.

Polarization terms The electric field created by the oxide ions polarizes the metal and this gives rise to an image charge term associated with an interaction between charges and induced dipoles. It should not be mistaken with the previous electrostatic term which is a charge–charge interaction. The image term has been discussed in detail by Stoneham and co-workers (Stoneham and Tasker, 1987; Finnis *et al.*, 1990). The example of a rocksalt{100} surface in contact with a metal shows that σ_{pol} is of the same order of magnitude as W_{adh}. If the metal acts as a perfect electrostatic mirror, and if the mirror plane is equidistant from two atomic planes, the electrostatic images are located at the positions of the missing oxide lattice sites (Fig. 5.12). Thus, σ_{pol} is equal to the Madelung surface energy. Along the {110} orientation this is no longer the case, but σ_{pol} remains equal to a fraction of the Madelung surface energy.

This image model presents several drawbacks. It yields the same value of σ_{pol} whatever the metal, contrary to the experiments. In addition, it predicts σ_{pol} values which are strongly dependent upon the location of the interface and upon the ionic charges, parameters which are not precisely known. Nevertheless, in agreement with experiments on irradiated mate-

rials, it shows that the adhesion is improved by the presence of charged
defects, which are attracted by their electrostatic images (Tasker and
Stoneham, 1987; Duffy *et al.*, 1993; 1994). However, defects do not necessarily induce a better adhesion. Contact angle measurements of liquid
gold deposited on sub-stoichiometric titanium oxides TiO_{2-x} have proved
that an increase of x of more than two orders of magnitude is associated
with no significant variation in the wetting angle (Chabert, 1992).

Exchange and correlation energy

Correlations include all the electron–electron interaction processes which
are neglected in the Hartree–Fock mean field approximation. Their contribution to the interfacial energy comes from the changes in these processes
which take place when the interface is formed.

At a metal–vacuum interface the most important contribution comes
from the electrons located outside the metal, which are less screened than
in the bulk. Their interaction with the correlation hole left behind them
yields a positive term to the surface energy – typically 1.46 J/m^2 for a
magnesium–vacuum interface, in the jellium approximation.

At a metal–insulator interface, the correlation term arises from the
mutual polarization of the two media, i.e. from the dispersion forces.
Naidich (1981) has assumed that, besides a small chemical term, the
adhesion energy involves mainly van der Waals pair interactions between
atoms located in the close vicinity of the interface. By analogy with
London's formula, he has written W_{adh} in the following way:

$$W_{adh} = \frac{3}{2} \frac{\alpha'_M \alpha'_I}{R^6} \left(\frac{I_M - I_O}{I_M + I_O} \right) , \qquad (5.6.2)$$

as a function of the metal atom and O^{--} polarizabilities: α'_M and α'_O – the
cation polarizability is neglected – ; and as a function of the ionization
potentials of these two species: I_M and I_O. R denotes the mean inter-
atomic distance across the interface. The energies predicted by (5.6.2)
have the right order of magnitude compared to the experimental adhesion
energies, but Eustathopoulos *et al.* (1991) have noticed that the variations
from one system to another are too low, compared with the experimental
results.

A possible reason for this discrepancy is the neglect of the band polarizability of both media, which involves virtual excitations across the
insulator gap and across the metal Fermi level. Barrera and Duke (1976)
have accounted for these excitations in a model similar to the one used
by Inglesfield and Wikborg (1975) and Wikborg and Inglesfield (1977) for
metal–metal van der Waals interactions. The estimated van der Waals

energy σ_{VW} at an interface between two media 1 and 2 then reads:

$$\sigma_{VW} = \frac{\hbar}{2}\int_0^{e^2}\frac{\mathrm{d}g}{g}\int\frac{\mathrm{d}^2q}{(2\pi)^2}\int\frac{\mathrm{d}\omega}{(2\pi)}\Im\frac{[\epsilon_1(\omega)-\epsilon_2(\omega)]^2}{\epsilon_1(\omega)\epsilon_2(\omega)[\epsilon_1(\omega)+\epsilon_2(\omega)]}.\quad (5.6.3)$$

This expression involves three integrations: on the coupling strength g, on the component q of the wave vector parallel to the surface and on the frequencies ω. The integral over ω yields three terms associated with (i) the bulk plasmon poles ω_{bn}, for which the $\epsilon_n(\omega)$ vanish ($n=1$, 2); (ii) the bulk transverse excitations ω_{Tn}, for which $\epsilon_n^{-1}(\omega)$ vanish; and (iii) the two interfacial frequencies ω_{in}, for which $\epsilon_1(\omega)+\epsilon_2(\omega)$ vanishes:

$$\sigma_{VW} = \frac{\hbar}{4}\int\frac{\mathrm{d}^2q}{(2\pi)^2}\left(2\sum_n\omega_{in}-\omega_{b1}-\omega_{b2}-\omega_{T1}-\omega_{T2}\right).\quad (5.6.4)$$

σ_{VW} is thus proportional to the change in the plasmon zero-point energy.

Assuming that the macroscopic ($q = 0$) dielectric constants of the insulator and the metal, respectively, are equal to:

$$\epsilon_I(\omega) = 1 + \frac{\hbar^2\omega_{pI}^2}{E_g^2 - \hbar^2\omega^2 - i\hbar^2\omega/\tau},\quad (5.6.5)$$

(Equation (4.2.15)), and:

$$\epsilon_M(\omega) = 1 - \frac{\omega_{pM}^2}{\omega^2 + i\omega/\tau},\quad (5.6.6)$$

one can show that (5.6.6) yields the right order of magnitude for the adhesion energy and allows a discussion of its variations with the metal electronic density, which fixes ω_{pM}, and with the insulator gap-width Δ (Didier and Jupille, 1994b). As the insulator ionicity grows, it is found that σ_{VW} decreases. Stoneham (1983) has used Barrera and Duke's model to derive a simple wetting criterion, relying upon a comparison between the metal plasmon frequency and a critical frequency value:

$$\hbar\omega_c = 2\sqrt{2}\left(\sqrt{\hbar^2\omega_{pI}^2 + 2\Delta^2} - 2\Delta\right).\quad (5.6.7)$$

In this expression they have assumed that $E_g = \Delta$ (for a discussion, see Section 4.2). Metals with low plasmon energies can wet ($\theta < \pi/2$) narrow-gap insulators: $\omega_{pM} < \omega_c$. Conversely, no wetting occurs when $\Delta > \hbar\omega_{pI}/\sqrt{2}$. Two families of metals can thus be distinguished depending on the values of their plasmon energies. The noble metals and some transition metals, such as chromium, have high plasmon frequencies ω_{pM} – $\hbar\omega_{pM}$ typically larger than 25 eV. Other transition metals together with lead, tin and aluminium are in an intermediate range. Alkaline and alkaline-earth metals have the lowest plasmon frequencies, $\hbar\omega_{pM}$ less than

10 eV. As far as insulators are concerned, the ratio $\Delta/\hbar\omega_{\rm pI}$ is related to the optical dielectric constant ϵ_∞ by: $\Delta/\hbar\omega_{\rm pI} = 1/\sqrt{\epsilon_\infty - 1}$ (Equation (5.6.5) when $\Delta = E_g$). Its value distinguishes three families: the alkali-halides, the oxides, and the covalent insulators, associated respectively with $\Delta/\hbar\omega_{\rm pI}$ of the order of 0.6 to 0.5, 0.5 to 0.3 and 0.3 to 0.1. Using experimental values of the wetting angles, Stoneham has tried to classify the interfaces according to the criterion (5.6.7). He has concluded that the van der Waals energy alone, estimated in the continuous approach of Barrera and Duke, cannot produce a discrimination between wetting and non-wetting interfaces in agreement with experiment. He has proposed another classification, based upon the value of the optical dielectric constant, which shows that non-reactive liquid metals wet substrates with a large refractive index (ϵ_∞ larger than 4.5). This criterion may also be used to understand metal–support interaction in catalysis, a strong interaction with the support corresponding to good wetting.

In the present state of the literature on metal–oxide interfaces, Barrera and Duke's approach seems the best treatment of van der Waals interactions, but its results should not be directly compared with experimental adhesion energies, since it represents a single contribution. This point is also true for the image term.

5.7 Conclusion

Macroscopic trends in metal–oxide interfacial phenomena seem now well established. However, a detailed microscopic understanding is still lacking on several points, such as the atomic energy level positions in the vicinity of the interface, the interfacial metal–oxygen or metal–cation charge transfers and the precise role of defects – be they point defects, such as the oxygen vacancies, or extended defects, such as the dislocation network which reduces the elastic stress associated with the lattice misfit. A study of well-controlled model systems may help in answering these questions in the coming years. It is only when they are solved, both experimentally and theoretically, that one will be able to actually control the deposition of metal atoms and the formation of good metallic films on oxide surfaces, which is an important challenge in several areas of applications.

6

Acid–base properties

The concept of acid–base interaction is often used in the physics of adhesion to account for the formation of interfacial bonds, in heterogeneous catalysis to analyze the reactivity and selectivity of solid catalysts, and in colloid physics to describe the double-layer interactions between small particles in suspension in a liquid. This chapter will present a brief historical sketch of the acid–base concept in molecular physics and will specify its relevance in adhesion and catalysis. Then, the acid–base characteristics of oxide surfaces will be described, with special attention paid to the properties of hydroxyl groups on these surfaces.

6.1 Historical background

Although the existence of acids had been recognized since the Middle Ages, it was only in the 17th century that the antagonism between acids – from the latin 'acetum' – and bases – from the arabic 'al kalja', which became alkali – became clear. A base is defined as an anti-acid, which reacts in an effervescent way with an acid to produce a salt (Finston and Rychtman, 1982). In 1815, Davy and Dulong postulated that hydrogen is a necessary constituent of acids.

From 1880 onwards, the specific properties of bases were studied. Realizing that reactions in solutions involve ionized species, Arrhenius defined an acid as an hydrogen-containing substance that dissociates into a proton H^+ and an anion when dissolved in water:

$$XH \rightarrow H^+ + X^- , \qquad (6.1.1)$$

and a base as a substance which dissociates into an hydroxyl group OH^- and a cation under the same circumstances:

$$XOH \rightarrow OH^- + X^+ . \qquad (6.1.2)$$

160

Salts, which are the products of acid–base reactions, are thus not necessarily electrically neutral but rather exhibit well-defined acid–base properties in water. Acidity appears as a relative concept and the existence of amphoteric substances, i.e. substances which can behave either as acids or bases depending on the circumstances, is stressed. The proton concentration, $[H^+]$, quantifies the acidity of a solution and the pH scale is established:

$$pH = -\log[H^+] \,. \tag{6.1.3}$$

A small pH indicates an acidic solution and a large pH indicates a basic solution. Yet, Arrhenius' definition presents several shortcomings. It does not recognize as acids molecules such as SO_2 or CO_2 which do not contain hydrogen atoms, nor as bases molecules like NH_3 without OH groups. It restricts the acid–base reactions to the aqueous medium and postulates the existence of the free proton H^+.

The next historical stage is the Brønsted–Lowry theory (Brønsted, 1923; Lowry, 1923a; 1923b) which generalizes the definition of a base to proton-acceptor substances:

$$Acid \iff Base + H^+ \,. \tag{6.1.4}$$

Each acid is now associated with a base. An amphoteric molecule may belong to several conjugate acid–base pairs, and may serve as the acid in one conjugate pair and as the base in another pair. Brønsted denies that acid–base reactions necessarily produce salt and water. He proposes the more general protolysis equation, independent of the solvent:

$$Acid_1 + Base_2 \iff Base_1 + Acid_2 \,. \tag{6.1.5}$$

Reactions with coloured indicators are thus recognized to be genuine acid–base reactions. By discarding the aqueous reference point, Brønsted claims that no universal point of neutrality exists and that acid and base strengths vary in opposite directions along the same continuum. The relativity of acidity scales is established. It is interesting to note that, according to (6.1.4), at least one member of every conjugate acid–base pair is an ion. This remark suggests that the acidity of a substance may be reinforced by increasing its positive charge, because this favours the proton separation, or by decreasing the formal negative charge of its central anion, in the case of hydrogenated molecules such as CH_4, NH_3, OH_2 and FH.

Brønsted's definition acknowledges the existence of acid–base reactions in the gas phase, in which the complicating effects of solvation are absent. The acidity of a substance B is then measured by the proton affinity PA of the molecule, i.e. the variation of enthalpy involved in the reaction:

$$B + H^+ \iff BH^+ \,. \tag{6.1.6}$$

An increasing basicity is associated with an increasing PA. The proton affinity may be expressed as a function of the energy of homolytic dissociation D of BH, the electronic affinity EA of B and the ionization potential I of hydrogen:

$$PA = D - EA + I \, . \tag{6.1.7}$$

However, PA is not an easily measurable quantity, because of the rotational and vibrational degrees of freedom of the molecules. [A correlation has recently been established between the oxygen 1s core-level shift and the oxygen proton affinity, in oxygen-containing molecules (Martin and Shirley, 1974; Davis and Rabelais, 1974; Davis and Shirley, 1976).]

Lewis (1923) subsequently stressed that ions play too important a role in the older theories, and that one should not restrict the definition of an acid–base pair to proton-containing substances. He redefined a base as a molecule with a free lone electron pair available to complete the stable electronic configuration of an acidic molecule. In a symmetric way, an acid may accept a lone electron pair to complete its own stable configuration. In other words, a base is an electron pair donor, and an acid an electron pair acceptor. In Lewis' picture in which ':' represents an electronic doublet, an acid–base reaction reads:

$$A + :B \iff A:B \, . \tag{6.1.8}$$

This new definition includes the reactions which take place in non-aqueous solvents or on solids. Lewis insisted that there is no universal acidity scale. A change of reference substance may not only alter but may sometimes completely invert orders of strength. This happens because a reaction energy is fixed not only by the characteristics of the reactants but also by factors which are functions of the final product: geometric configurations, number of bonds, etc.

Mulliken (1951; 1952a; 1952b) and Hudson and Klopman (Hudson and Klopman, 1967; Klopman and Hudson, 1967) rephrased Lewis' electronic theory with the tools of quantum mechanics. Mulliken wrote down the wave function of an adduct AB, produced by a reaction between A and B, in the following form:

$$\Psi_{AB} = a\Psi(A, B) + b\Psi(A^-, B^+) \, . \tag{6.1.9}$$

$\Psi(A, B)$ is a 'no-bond' wave function, in which no electron transfer has taken place. It accounts for all electrostatic interactions between A and B. $\Psi(A^-, B^+)$ is the wave function after complete transfer of one electron from B to A. The degree of charge transfer is determined by the ratio b^2/a^2, which may take any value between zero – highly ionic bond – and unity – covalent bond. When b^2/a^2 is small, the energy associated with

Ψ_{AB} is equal to:

$$E_{AB} = E_0 - \frac{2(\beta_{12} - E_0 S_{12})^2}{E_1 - E_2} . \qquad (6.1.10)$$

E_0 corresponds to $\Psi(A, B)$. It is a function of the net charge densities on the donor and acceptor sites. The second term results from covalent effects. β_{12} and S_{12} are the resonance and overlap integrals, respectively, along the AB bond and, to a good approximation, the excitation energy $E_1 - E_2$ can be taken equal to the energy difference between the base ionization potential and the acid electron affinity.

In a closely related way, Klopman and Hudson expanded the adduct wave function in terms of the reactant states. They wrote the energy involved in the adduct formation as:

$$\Delta E = -\frac{Q_1 Q_2}{\epsilon R_{12}} + 2 \sum_m \sum_n \frac{|c_1^m c_2^n \beta_{12}|^2}{E_m - E_n} . \qquad (6.1.11)$$

The first term is the screened electrostatic interaction between the donor and acceptor charges Q_1 and Q_2 – assumed to be point charges – at the equilibrium distance R_{12} in the adduct. The second term accounts for covalent effects. The factor of 2 indicates that two electrons are shared. The c coefficients are the molecular wave-function weights on atoms 1 and 2. The energies E_m and E_n are equal, to a first approximation, to the frontier orbital energies: the base HOMO (HOMO = highest occupied molecular orbital) and the acid LUMO (LUMO = lowest unoccupied molecular orbital), i.e. to the base first ionization potential and to the acid electron affinity. A typical frontier orbital diagram is shown in Fig. 6.1.

When the covalent term is negligible, i.e. when the reaction involves high-charge-density atoms at small inter-atomic distances, the reaction is charge-controlled. In the opposite limit – atoms with large atomic radii and nearly degenerate base HOMO and acid LUMO – the reaction is frontier-controlled, and the electron transfer between the reactants is large.

In the Mulliken, Hudson and Klopman quantum treatments, the degree of charge transfer may take any value. Consequently, they apply not only to Lewis-type acid–base reactions, in which, strictly speaking, an electron pair is shared, but also to oxido–reduction reactions:

$$Red \iff Ox + e^- , \qquad (6.1.12)$$

characterized by a single electron exchange. In the modern theories of chemical reactivity (Usanovich, 1939; Jensen, 1978; 1980), the oxidizing power, the acidity, the basicity and the reducing power are steps along a continuum, rather than distinct phenomena.

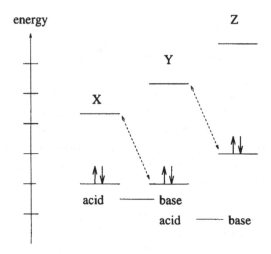

Fig. 6.1. Frontier orbital diagram for an acid–base reaction: the HOMOs are filled with an electron pair; the LUMOs are empty. The Y compound presents an amphoteric character: it is a base in the XY conjugate pair, and an acid in the YZ pair. When the energy difference between the base HOMO and the acid LUMO is small, the reaction is controlled by the frontier orbitals. Otherwise, it is charge-controlled.

After Mulliken, the most important steps in the development of the acid–base concept, are Pearson's HSAB principle (1963; 1966) (HSAB: Hard Soft Acid Base) and Drago and Wayland's (1965) $\mathscr{E}\&\mathscr{C}$ equations (see below). Pearson distinguished two types of bases: the hard bases with low-energy HOMOs and small and weakly polarizable atoms, and the soft bases with highly polarizable atoms of low electronegativity. A similar distinction exists between soft and hard acids. Hardness and softness are relative qualities, with an ill-defined border line. Hard acids prefer to react with hard bases and soft acids with soft bases, and the reaction speeds are high. Myers (1974) criticized the postulate which underlies the HSAB theory. According to him, a high polarizability does not necessarily induce a high covalency. The latter is due rather to a low difference in electronegativity between the acid and the base.

Parr and Pearson (1983) defined the hardness of a molecule as a function of its total energy E and its electron number N. The first partial derivative of E with respect to N is the Mulliken electronegativity χ:

$$\chi = -\frac{\partial E}{\partial N}, \tag{6.1.13}$$

and the second derivative is the absolute hardness η:

$$\eta = \frac{1}{2}\frac{\partial^2 E}{\partial N^2}. \tag{6.1.14}$$

These quantities are easily estimated in a density functional approach. The concept has been enlarged by Yang and Parr (1985) and Parr and Yang (1986) to describe local properties of large molecules and, thus, predict the more favourable reaction points.

To a good approximation, χ and η may be expressed as a function of the ionization potential I and the electron affinity A of the molecule, by $\chi = (I + A)/2$ and $\eta = (I - A)/2$. In a one-electron picture, they are proportional to the sum and difference of the HOMO and LUMO energies (Pearson, 1989). The relationship between the hardness and the gap width $E_{LUMO} - E_{HOMO}$ allows a simple transposition of the concepts of molecular chemistry to the solid state.

Drago and Wayland (1965) proposed an empirical equation which expresses the formation enthalpy of an adduct AB in terms of two parameters $\mathscr{E}$ and $\mathscr{C}$ characterizing the electrostatic and covalent strengths of its constituents:

$$-\Delta H = \mathscr{E}_A \mathscr{E}_B + \mathscr{C}_A \mathscr{C}_B . \qquad (6.1.15)$$

This equation is consistent with the iono–covalent concept. Reactions involving acids and bases with large $\mathscr{E}$ values are charge-controlled, while high-$\mathscr{C}$ reactants yield frontier-controlled reactions. Nevertheless, a given species may have both a high $\mathscr{E}$ and a high $\mathscr{C}$. Electrostatic and covalent interactions are not mutually exclusive. Drago and Wayland's equation applies only to neutral species and implicitly assumes a small degree of charge transfers. The $\mathscr{E} \& \mathscr{C}$ equation is the empirical descendant of Mulliken's quantum mechanical formulation. The classification of acids and bases relies on the arbitrary choice of iodine as a reference point ($\mathscr{E} = 1$, $\mathscr{C} = 1$). $\mathscr{E}$ and $\mathscr{C}$ values have been compiled for many acids and bases, and account for the formation enthalpy of a large number of adducts. Despite their names, it is not obvious that the parameters actually reflect the electrostatic and covalent parts of the interactions.

In the context of molecular chemistry, the acid–base concept has, thus, been progressively refined and enlarged, until it merged into the electron donor and acceptor concept. It is now well established that assigning a value to the acid–base strength requires a reference point and that scales based upon different references may present strong deviations. Such acidity scales are nevertheless useful in well-defined contexts, and the reader is referred to review papers giving their specificities and limitations (Jensen, 1991). On fundamental grounds, one has nevertheless to keep in mind that the variation of enthalpy in an acid–base reaction is fixed not only by the intrinsic properties of the reactants – polarizability, size, frontier orbitals, etc., – but also by factors which are characteristics of the adduct – relative conformation of the two molecules in the adduct, number of bonds formed, etc., – as previously recognized by Lewis.

When a solid takes part in an acid–base reaction, new degrees of
freedom have to be taken into account, for example the surface orientation,
the site coordination number, the structural or stoichiometry defects,
some steric factors, etc. For a surface, as well as for a large molecule,
it is no longer possible to talk of *one* acid–base strength. One has
to distinguish between non-equivalent surface sites and quantify their
reactivity. The latter is not determined by local factors as in small
molecules, because long-range electrostatic and covalent effects occur. For
example, on an insulating surface, the Madelung electrostatic potential
acts not only on the substrate atoms but also on the reacting molecules.
The substrate electronic states involved in the reaction — equivalent
to the frontier orbitals — are delocalized in the whole solid. Finally,
the electrostatic perturbations produced by the reacting molecules induce
electron redistributions in the solid, on length scales much larger than
the inter-atomic distances. The actual importance of such effects on the
surface reactivity is, at the present time, not well known. In the next two
sections, we will specify how the acid–base concept applies to reactions
or adsorption on oxide surfaces, in the context of adhesion science and
catalysis.

6.2 The acid–base concept in adhesion science

The adhesion between two materials is the manifestation of the forces at
work across the interface. A high level of adhesion is sought in joints,
coatings or at the metal–oxide interfaces discussed in the previous chapter.
It is of special importance, due to economical implications, in the glue
industry – mechanical joints are more and more often being replaced by,
or used in parallel with, adhesive joints – and in the paint and glass
industries.

Two parameters characterize the quality of a joint: its resistance to
separation and its resistance to ageing. Reactions, such as the hydrolysis
of interfacial bonds, induced by external agents, are often responsible for
premature ageing (Ishida and Koenig, 1980; Gan *et al.*, 1986; Stein and
Prutzman, 1988; Brewis and Kinloch, 1983). The resistance to separation,
on the other hand, is measured by various peeling tests. It depends upon
several factors:

- The contact specific area: mechanical adhesion may be improved by
  increasing the surface roughness or porosity, but this parameter alone
  cannot account for the measured failure strengths.
- The existence of an inter-diffusion zone at the interface, called an
  interphase, may also favour adhesion. We have already met this
  phenomenon when discussing metal–oxide interfaces. When a reactive wetting induces atomic displacements, the wetting angle is small

and the adhesion energy large. The same is true for other materials, for example in the quickly developing field of metal–polymer interfaces.

• In the absence of interdiffusion, the failure strength is generally written as the product of two terms: the reversible work of adhesion, determined by the interfacial properties of both species, and a factor of energy dissipation resulting from irreversible deformations during the failure (Gent and Schultz, 1972; Schultz and Gent, 1973; Carré and Schultz, 1984). While the first term ranges from 0.1 to 1 J/m^2, according to whether van der Waals or covalent forces drive the interfacial bonding, the second term may be 10^3 to 10^4 times larger. It strongly depends upon the peel-test geometry. We will not comment on this factor, which is the subject of much research in failure mechanics (Gillis and Gilman, 1964; Kendall 1971; 1975; Andrews, 1974; Maugis, 1977) and in polymer science (Lake and Thomas, 1967; de Gennes, 1989; Brown, 1993).

Acid–base interactions are involved in the first factor. It is recognized, for example, that the interfacial bonding at a solid–polymer interface has two contributions: one due to the van der Waals (or dispersion) forces and one, W_{AB}, due to acid–base interactions. W_{AB} is proportional to the number n of reactive surface sites and to the enthalpy change ΔH_{AB} per formed bond. Fowkes (1987) stressed the role of acid–base interactions in the adsorption of basic (PMMA) or acidic (PVC) polymers on silica (acidic) or calcium carbonate (basic) surfaces. The quantities n and ΔH_{AB} may be determined by gas phase chromatography, ellipsometry, infra-red spectroscopy, NMR (NMR= Nuclear Magnetic Resonance) and x-ray photoemission (Fowkes, 1990). Such studies aim at understanding which surface modification may yield stronger interfacial forces and better adhesion (Mittal and Anderson, 1991).

Under usual conditions of application, surfaces present a well-defined Brønsted acidity or basicity. On contact with ambient atmosphere, a surface oxidizes and gets covered with several layers of water. The molecules in contact with the substrate are dissociated into protons and hydroxyl groups, which strongly bind to the oxide ions. The outer layers are made of non-dissociated molecules which interact between themselves and with the dissociated species by hydrogen bonding. Such a description, sketched in Fig. 6.2, is supported by adsorption heat (Q) measurements. During the first stages of hydration, Q ranges from 125 kJ/mol to 300 kJ/mol, while in the next stages, it decreases down to 60–80 kJ/mol – the typical binding energy between a water molecule and an hydroxyl group – and then to 40 kJ/mol – the hydrogen bond strength between two water molecules (Texter *et al.*, 1978).

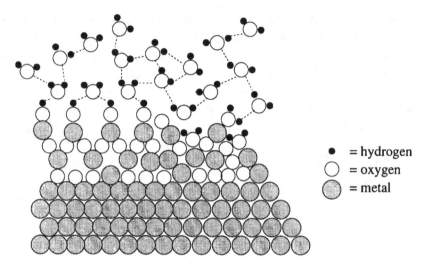

Fig. 6.2. Representation of a metallic surface in contact with an oxidizing and humid atmosphere. Point lines between oxygens and hydrogens in the hydroxylation layer indicate hydrogen bonds.

An hydroxylated surface may, thus, be a proton donor (Brønsted acid):

$$MOH \rightarrow MO^- + H^+ ,\qquad(6.2.1)$$

or a proton acceptor (Brønsted base):

$$MOH + H^+ \rightarrow MOH_2^+ .\qquad(6.2.2)$$

An acidic pK_1 and a basic pK_2 are associated with the equilibrium constants K_1 and K_2 of these two reactions. Electrochemical measurements allow a determination of the isoelectric point of the surface ($IEPS$) which is the pH value of a solution in contact with the surface which yields an equal concentration of MOH_2^+ and MO^- sites. The surface then bears a net zero charge and the proton concentration is equal to:

$$[H^+]^2 = \frac{K_1}{K_2} .\qquad(6.2.3)$$

The $IEPS$ may also be expressed as:

$$IEPS = \frac{1}{2}\left(\frac{-\Delta G^0}{2.3RT}\right) ,\qquad(6.2.4)$$

as a function of the free energy change ΔG^0 in the total reaction:

$$MO^- + 2H^+ \rightarrow MOH_2^+ .\qquad(6.2.5)$$

A high $IEPS$ value reveals a strong surface basicity, while a value close to zero is characteristic of a strong acidity. Parks (1965) compiled $IEPS$

values for many oxides. The highest values are found in alkaline or alkaline-earth metal oxides, and the lowest in oxides of M_2O_5 or MO_3 stoi-chiometries. The $IEPS$ are generally measured on powders, although, more recently, some determinations have been performed on thin supported films (Lin and Arribart, 1993).

Bolger (1983) proposed a model which accounts for acid–base interactions between a polar polymer and an oxide surface. He characterizes a polymer by its dissociation constant, $K_A = [H^+][X^-]/[HX]$ in the case of an acid:

$$XH + H_2O \rightarrow H_3O^+ + X^-, \qquad (6.2.6)$$

or $K_B = [H^+][X]/[HX^+]$ in the case of a base:

$$X + H_3O^+ \rightarrow HX^+ + H_2O, \qquad (6.2.7)$$

and the oxide surface by its isoelectric point $IEPS$. After a number of simplifications, the adhesion energy is found to be proportional to $W_A = IEPS - pK_A$ or $W_B = pK_B - IEPS$, depending upon the polymer characteristics. When interfacial dispersion forces prevail, W is negative. When acid–base interactions take place, large and positive W are induced. Ultimately, a chemical attack of the solid or corrosion may result. The strongest adhesion thus takes place, for example, between a silica substrate (acid) and an amine-based adhesive (strong base), between an MgO substrate (base) and an adhesive containing carboxyl (strong acid) or phenol groups (weak acid), or between an amphoteric substrate (aluminium or iron oxide) and either type of adhesive.

According to Bolger, the adhesion energy is, thus, proportional to the difference between two quantities which characterize the two reactants. The degrees of freedom of the adduct *surface + adhesive*, such as the relative geometry of the surface and the polymer, modifications in the electronic states of one species in the presence of the other, etc., are neglected. This appears clearly by formally dividing the adhesion process into three steps:

- the acidic dissociation of the polymer:

$$HX \rightarrow H^+ + X^-, \qquad (6.2.8)$$

- the proton capture by the surface:

$$MOH + H^+ \rightarrow MOH_2^+, \qquad (6.2.9)$$

- the adsorption of X^- on the surface:

$$MOH_2^+ + X^- \rightarrow MOH_2^+ \ldots X^-. \qquad (6.2.10)$$

To be specific, we have written these equations for an acidic polymer. To obtain Bolger's result, it is necessary to assume that the dissociation constants K_1 and K_2 have symmetric behaviours around $IEPS$ ($K_2 = 1/K_1 = 1/10^{-IEPS}$) and to neglect the adsorption energy of X^- on the surface (6.2.10). The latter assumption can only be justified when interfacial forces are weak, i.e. when the adhesion is driven by dispersion forces rather than acid–base forces.

Precise polymer conformations at interfaces and adhesive–substrate charge transfers are not well known, at the present time. But, as we will see in Section 6.5, the acidity characterization of surface sites in oxides is more advanced.

6.3 The acid–base concept in heterogeneous catalysis

Solid acid catalysts have been used for many years in petroleum chemistry and organic syntheses. More recently, several types of solid bases have been discovered and used in heterogeneous catalysis (Tanabe, 1970; 1981). Oxides occupy an important place among the catalysts, due to the wide range of their structural and electronic properties. This is true for simple oxides like Al_2O_3, TiO_2, ZrO_2, V_2O_5 or MoO_3, more complex oxides such as natural clay minerals, bentonite, hetero-poly-anions and zeolites, and mixed oxides such as SiO_2–Al_2O_3, TiO_2–MoO_3, etc.

It is recognized that the use of solid catalysts presents several advantages over their liquid homologues. They frequently have a high activity and selectivity. They do not corrode the vessels or the reactors, and induce no environmental pollution. It is easy to separate them from the reaction products and, thus, to make repeated use of them. Due to their industrial interest, they have been the subject of many studies over the last thirty years.

Various characterization methods have been developed to measure the number, the nature and the strength of acid and basic sites at the surface of solid catalysts. They include for example the adsorption of indicators on surfaces and titration. Catalysts are transformed into powders in a test tube and successively mixed with various indicators in a non-polar solvent, in order to perform comparative measurements. The number of reactive sites with a given strength is then determined by amine titration for acidic sites or benzoic acid titration for basic sites. The acid strength H_0 is expressed by the pK_A values of the conjugate acids of basic indicators, while the base strength is given by the H_0 of the conjugate acid sites. Acidity and basicity are, thus, obtained on a common scale. It has been found that the highest H_0 values for acid and basic sites generally coincide. This common value $H_{0,max}$ is, thus, the relevant parameter representing the acid–base character of solid surfaces (Yamanaka and Tanabe, 1975;

1976). A solid with a high positive $H_{0,max}$ has strong basic sites and weak acidic sites. In a symmetric way, the presence of strong acidic sites and weak basic sites is associated with a high negative $H_{0,max}$. There exists a qualitative correlation between $H_{0,max}$ and the isoelectric point *IEPS* although the acid–base reactions relevant for these two quantities do not take place in the same solvents – non-polar solvents for $H_{0,max}$ and aqueous solutions for *IEPS*.

The thermo-desorption of molecules in the gas phase also yields information on the surface acidity. Since desorption of a base from a strongly acidic site is more difficult than from a weakly acidic one, the number of desorbed molecules as a function of temperature gives an estimate of the strength of the reactive sites. The test molecules which are commonly used include ammonia, pyridin and amines to quantify the surface acidity, and CO_2, nitric acid and phenol to quantify their basicity.

The technique of isotherm microcalorimetry allows a simultaneous determination of the strength and energy distribution of the adsorption sites (Auroux and Gervasini, 1990). Fig. 6.3 displays several possibilities of adsorption for NH_3 on oxide surfaces. On rough surfaces, each of them may take place on non-equivalent surface sites: facets, step edges, kinks, etc., and, as a consequence, may be associated with a non-unique adsorption energy. One realises how difficult it is to determine which site is responsible for a measured energy. Planar single-crystal surfaces are not used in catalysis, because their specific area is too low and because they are less reactive than powders. However, once modified by a controlled amount of defects, they could allow a precise assignment of the reactive sites at a microscopic level. This strategy was used, for example, to understand the chemisorption of CO on single-crystal ZnO faces (Solomon *et al.*, 1993).

The measurement of the activity in a specific catalytic reaction, such as isopropanol decomposition or butene isomerization is also often used in the field. Catalysts may, for example, be characterized by their ability to dehydrate isopropanol molecules into propene, or to dehydrogenate them in acetone. The former reaction tests their acidity and the latter their basicity (Ai, 1977; Cunningham *et al.*, 1981; Nollery and Ritter, 1984; Tanabe *et al.*, 1989; Gervasini and Auroux, 1991).

The position of the oxygen 1s core levels in oxides was shown to be correlated with basicity. Oxides with weakly basic oxygens have their 1s core levels shifted towards higher binding energies. According to Vinek *et al.* (1977), the order of basic strengths determined by this method is the following: La_2O_3 (529 eV) > Sm_2O_3 (529.2 eV) > CeO_2 (529.4 eV) = Dy_2O_3 (529.4 eV) > Y_2O_3 (529.5 eV) > Fe_2O_3 (530.3 eV) > Al_2O_3 (531.8 eV) > GeO_2 (532.4 eV); P_2O_5 (532.4 eV) > SiO_2 (533.1 eV).

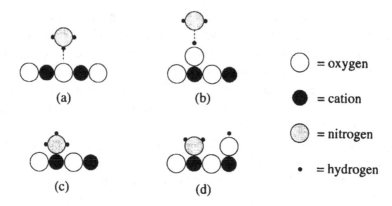

Fig. 6.3. NH_3 adsorption on oxide surfaces: (a) hydrogen bonding; (b) proton transfer; (c) Lewis' acid–base interaction; and (d) dissociative adsorption.

Values in brackets give the measured oxygen 1s-level binding energies. A correlation has been established between this energy and the *IEPS* in several oxides (Mullins and Averbach, 1988; Delamar, 1990; Casamassima *et al.*, 1991). However, some care must be taken, because the binding energies are also dependent upon electron relaxation around the 1s hole, not present in electrochemical measurements.

Finally, the values of the dielectric relaxation of hydroxyl groups measured by NMR (NMR= Nuclear Magnetic Resonance) and the frequencies of the vibrational modes of adsorbed molecules (OH^- or NH_3), recorded by infra-red absorption, are related to the adsorption strength. From experimental values of OH^- stretching frequencies and from the acidity scale relevant for acids in solutions, Hair and Hertl (1970) were, for example, able to determine the pK_A of adsorbed OH^- on surfaces. We will come back to the characteristics of hydroxyl groups adsorbed on oxide surfaces in the last section of this chapter.

In parallel to experimental approaches, empirical arguments have been used to correlate the acidity with some microscopic parameters of the oxides. We will now examine these parameters.

6.4 Qualitative discussion

One of the basic questions underlying many studies in the acid–base context is: 'Which are the specific properties of the reactants which determine the strength of an acid–base reaction?'. As far as oxides are concerned, depending upon the field of research, various physical parameters have been proposed: the cation electronegativity; the cation ionic radius and formal charge; the oxygen partial charge; and the surface site coordination. Their link with the surface acidity relies, in most cases, upon either very qualitative or empirical models. They will be presented in this section and their use will be rationalized on the basis of the theoretical models developed in previous chapters. Finally, some points related to mixed oxides will be considered.

Cation formal charge and ionic radius

In his compilation of the oxide isoelectric points, Parks (1965) showed a correlation between the $IEPS$ value and the cation characteristics: its formal ionic charge Q_M and its ionic radius r_M. As a function of Q_M, Parks was able to classify oxides according to decreasing $IEPS$:

$$M_2O \qquad Q_M = +1 \qquad 11.5 < IEPS$$
$$MO \qquad Q_M = +2 \qquad 8.5 < IEPS < 12.5$$
$$M_2O_3 \qquad Q_M = +3 \qquad 6.5 < IEPS < 10.4$$
$$MO_2 \qquad Q_M = +4 \qquad 0.5 < IEPS < 7.5$$
$$M_2O_5 \qquad Q_M = +5 \qquad IEPS < 0.5$$
$$MO_3 \qquad Q_M = +6 \qquad IEPS < 0.5 \,.$$

The basic oxides involve cations with the lowest formal charge. They are alkaline or alkaline-earth oxides. The cations in the acidic oxides have the highest formal charges: +5 or +6. The $IEPS$ values display a saturation value close to zero, because of the water levelling effect. Water, which is the solvent in $IEPS$ measurements, does not allow one to study species more acidic than H_3O^+, such as M_2O_5 and MO_3 oxides. Parks established that the $IEPS$ values are roughly linear decreasing functions of the ratio Q_M/r_M. Yet, the dependence on r_M implied by this relation is not systematically obeyed. There exist oxides of a given cation charge, in which the increase of $IEPS$ with r_M is well fulfilled:

$$MnO_2 \qquad r_{Mn} = 0.42 \text{ Å} \qquad IEPS \approx 4.2$$
$$TiO_2 \qquad r_{Ti} = 0.50 \text{ Å} \qquad IEPS = 4.7$$
$$SnO_2 \qquad r_{Sn} = 0.58 \text{ Å} \qquad IEPS = 7.3 \,.$$

But, in the following series, the trend is not obeyed:

$$MgO \qquad r_{Mg} = 0.65 \text{ Å} \qquad IEPS = 12.4$$
$$CdO \qquad r_{Cd} = 0.97 \text{ Å} \qquad IEPS = 10.4$$
$$HgO \qquad r_{Hg} = 1.10 \text{ Å} \qquad IEPS = 7.3 \,.$$

Similarly, in the context of heterogeneous catalysis, the average CO_2 adsorption energy on oxides was shown to increase, which reveals an

Fig. 6.4   Relative positions of the surface atoms in Parks' (1965) model.

increasing basic character, when the ratio Q_M/r_M diminishes (Auroux and Gervasini, 1990).

A simple electrostatic model accounts for the relationship between the acid character of the oxide and the ratio Q_M/r_M. Parks (1965) assumed that the free energy change ΔG^0, which determines the $IEPS$ value (Equation (6.2.4)) is mainly due to the work of the electrostatic forces when protons approach or leave the surface (Fig. 6.4). Assuming in addition that ions bear integer point charges, and that the interaction between the oxygen and the closest surface cation prevails, he deduced that:

$$\Delta G^0 = \frac{2Q_OQ_H}{\epsilon r_O} + \frac{2Q_MQ_H}{\epsilon(2r_O + r_M)} + \Delta G' , \qquad (6.4.1)$$

an expression in which $Q_O = -2$ and $Q_H = +1$; ϵ is the oxide optical dielectric constant, and r_O and r_M are the oxygen and cation ionic radii, respectively. The first term in (6.4.1) represents the proton attraction by the surface oxygen and the second term its repulsion by the neighbouring cation. $\Delta G'$ is the reaction free energy of non-electrostatic origin. Parks assumed that it remains constant whatever the oxide. Since only the proton–cation interaction distinguishes the various oxides in this model, $IEPS$ may be written in the most general form:

$$IEPS = A - B\frac{Q_M}{(2r_O + r_M)} . \qquad (6.4.2)$$

This equation accounts for the higher acid character of oxides involving cations with a high formal charge and a small ionic radius. Parks' electrostatic model may be criticized in three respects:

• First, it considers only short-range electrostatic interactions. Yet, in insulators, the electrostatic Madelung potential acting on a given atom results from an infinite sum of positive and negative contributions, and truncating the summation at the first-neighbour distance introduces large errors.

- The model neglects the bond formation between the adsorbate and the substrate. However, the O–H bond presents a strong covalent character and the covalent contribution to the bond energy varies from one oxide to another, as we will see in the next section.
- Finally, associated with the bond formation, an electron transfer takes place between the proton and the surface oxygens, which modifies the charges of both species. This charge modification has to be taken into account in the estimation of the electrostatic energy.

Parks' model predicts that oxide surfaces of different orientations have the same *IEPS*. Yet, Nabavi *et al.* (1993) showed, for example, that the pK_1 and pK_2 of nanometric-size particles of cerium oxide CeO_2 are very different on the (100), (110) and (111) faces. On the hydroxylated (100) face, oxygens with a coordination number $Z = 2$ are found, which belong either to the ideal surface or to adsorbed OH^- groups. On the two other faces, three-fold coordinated oxygens exist on the ideal surfaces, while singly coordinated oxygens result from the *on-top* adsorption of an OH^-.

Parks' electrostatic model was extended to account for the surface site environment, in the so-called MUSIC model (MUSIC = multisite complexation model) (Hiemstra *et al.*, 1989a; 1989b). As in Parks' model, the electrostatic forces are assumed to be responsible for the proton adsorption or desorption, but in order to take into account the oxygen environment, the model makes use of the bond charge and local charge neutralization concepts introduced by Pauling (1929; 1960). In an ionic crystal, the electric neutrality condition requires a balance between the cation charge and the anion charges in the neighbourhood. The neutralization may be described as a sharing of the positive cation charge between the neighbouring anions, with $q = Q_M/Z_b$ on each bond (Z_b is the cation coordination number in the bulk). This assumption is introduced in order to account, in an effective way, for the potential exerted by the ligands which surround the metallic ions. For example, it allows for the consideration of adsorbed species as isolated entities and, thus, avoids the estimation of the long-range Madelung field. The free energy associated with each elementary protonation reaction on a surface, then, takes a form very similar to that proposed by Parks:

$$\Delta G^0 = \frac{Q_O Q_H}{\epsilon R_{OH}} + Q_H \frac{Z q}{\epsilon R_{HM}} + \Delta G' . \qquad (6.4.3)$$

Here, Z is the oxygen coordination number and R_{OH} and R_{HM} are the proton–oxygen and proton–cation inter-atomic distances. By neglecting the proton radius, R_{HM} is approximated by the anion–cation inter-atomic distance in the oxide. The main difference between this expression and Parks' formula (6.4.1) lies in the value of the positive charge. Each of the

Z cations in the neighbourhood of the oxygen atom contributes as q. The pK associated with acidic or basic reactions thus depends in a linear way upon the oxygen coordination number Z:

$$pK = A - B\frac{Zq}{R_{HM}} .$$ (6.4.4)

The A and B constants are considered as empirical parameters and fitted to the behaviour of the corresponding molecules in a solution.

Applied to the cerium oxide particles, the MUSIC model predicts the following values for the elementary pKs:

$Z = 1$ $pK_1 = 24.16$ $pK_2 = 10.36$

$Z = 2$ $pK_1 = 14.8$ $pK_2 = 1.02$

$Z = 3$ $pK_1 = 5.48$ $pK_2 = -8.32.$

The basicity of the surface oxygens is, thus, larger and larger as their coordination number decreases (pK_1 variations), while the OH acidity is stronger as their coordination number increases (pK_2 variations).

Cation electronegativity and oxygen charge

Another parameter which is often invoked, when quantifying an oxide acidity, is the cation Mulliken electronegativity. It is defined as minus the first derivative of the atom energy E with respect to its electron number N: $\chi_M = -\partial E/\partial N$. Its relevance is easily understood since, according to the Lewis' definition, the acidity is the ability to accept an electron pair. The acidity is, thus, expected to be higher for cations of higher electronegativity. In a given oxide series, the higher the cation in the periodic table, the stronger the oxide acidity. The more on the left of the periodic table the cation is located, the more basic the oxide.

More precisely, it was recognized that it is not the *neutral* atom electronegativity which has to be considered, but rather the ionized atom electronegativity, considering its charge Q_M in the oxide. Following ideas developed by Iczkowski and Margrave (1961), Tanaka and Ozaki (1967) showed that, in first approximation, i.e. when assuming that the ion internal energy E is a quadratic function of the outer electron number N, the electronegativity reads:

$$\chi_M = \chi_{M0}(1 + 2aQ_M) .$$ (6.4.5)

For neutral atoms, it is equal to χ_{M0}, whose values have been compiled e.g. by Sanderson (1960) or Pauling (1929). In ions, χ_M varies linearly with Q_M. The coefficient $a = (3I_1 - I_2)/(I_2 - I_1)$ is related to the values of the first and second ionization potentials I_1 and I_2. It is close to unity in a large part of the periodic table.

Tanaka and Ozaki proved that the acid pK_A for the ionization of solvated cations:

$$M(H_2O)_m^{\alpha+} + H_2O \rightarrow [M(H_2O)_{m-1}OH]^{(\alpha-1)+} + H_3O^+ , \qquad (6.4.6)$$

varies monotonically with χ_M, and that the same is true for the *IEPS* of oxides, and for their catalytic activity in the reactions of propylene hydration, iso-butylene polymerization, and acetaldehyde polymerization. Similarly, Connell and Dumesic (1986a; 1986b) found that adding iron to a silica surface generates acid sites, the strongest of which are due to the presence of Fe^{3+} species and the weakest to Fe^{2+} species. They assigned this behaviour to the higher electronegativity of Fe^{3+} ions. It is not surprising to find a good correlation between the oxide *IEPS* and the cation electronegativity since, when comparing oxides with different stoichiometries, it is likely that the charge value Q_M induces the largest changes in χ_M. However, one should remember that the microscopic processes invoked in the electrostatic model and in Tanaka and Ozaki's approach are different. The first one considers only the electrostatic interactions, while the second relies on the ability of a cation to form a strong covalent bond.

In oxides, the ionicity of the oxygen–cation bond is strongly correlated with the cation electronegativity. As χ_M increases, the oxygen–cation bond becomes more covalent and the absolute value of the oxygen effective charge Q_O decreases. Sanderson (1964) has proposed an empirical method to estimate the partial ionic charges in a binary compound, which assumes that the charge transfer equalizes the electronegativities χ of the two species. With this argument, he predicted that $|Q_O|$ decreases in the series: Na_2O, MgO, Al_2O_3, SiO_2 and SnO_2. Tanabe and Fukuda (1974), by measuring the CO_2 adsorption energy in the series of alkaline-earth oxides, found that the basicity decreases from BaO to MgO: $BaO > SrO > CaO > MgO$. Basicity, thus, varies monotonically with the oxygen charge which, according to Sanderson, is equal to -0.61, -0.60, -0.57 and -0.50 in the series. This argument was used by Auroux and Gervasini (1990), who proved, for a large number of oxides, that the NH_3 adsorption energy decreases when the percentage of ionic character, defined according to Sanderson's scale (1960), increases, while the CO_2 adsorption energy increases along the same series. With this same criterion, Vinek *et al.* (1977) explained the results of the oxygen 1s core level shifts, quoted in the preceding section, and showed that an increase in the basicity is associated with a larger oxygen charge.

Sanderson (1964) also introduced the atom environment in his estimation of the effective charge. In the fluoride series, he showed that the fluorine partial charge Q_F decreases when the ion coordination number Z decreases. For example, $Q_F = -0.89$ in CsF ($Z = 6$), $Q_F = -0.56$ in BaF_2

$(Z = 4)$, $Q_F = -0.29$ in BeF_2 ($Z = 2$) and $Q_F = -0.17$ in SiF_4 ($Z = 1$). A similar result applies to oxides.

First discussion of these parameters

In the first chapters of this book, the electronic structure of oxides in the bulk and at the surface *in the absence of adsorbates* has been analyzed, and some theoretical models were given as guidelines for their understanding. It is possible to use them now to point out how the arguments of electronegativity and partial charge developed above are related.

Under the most simplifying Hartree approximation, when the eigenstates of the Hamiltonian are developed on an atomic orbital basis set, the diagonal terms of the Hamiltonian matrix, which represent effective renormalized atomic orbital energies, read (in atomic units):

$$\epsilon_i = \epsilon_i^0 - U_i Q_i - V_i , \qquad (6.4.7)$$

with ϵ_i^0 ($i = $ M, O) the atomic energies of the neutral atoms, $-U_i Q_i$ the intra-atomic correction associated with the excess (on the oxygens) or loss (on the cations) of electron–electron repulsion (U_i the intra-atomic electron–electron repulsion integral) and V_i the Madelung potential exerted on atom i by all other ions. This expression may be used to estimate the ion internal energy E_i and to write down the Mulliken electronegativity $\chi_i = -\partial E_i / \partial N_i$. Assuming that a single outer atomic orbital is involved in the chemical bond, χ_i reads:

$$\chi_i = \chi_{i0} + U_i Q_i + V_i . \qquad (6.4.8)$$

As in (6.4.5), this expression of χ_i contains a corrective term due to intra-atomic electron–electron interactions, although, here, the effective charge rather than the formal charge has to be used. In addition, solid state effects give a contribution to χ_i, equal to the Madelung potential V_i. Equation (6.4.8), thus, gives a generalization of the concept of electronegativity, suited to processes which take place in a solid or on a surface. It accounts for the variations of electronegativity as a function of the charge state and as a function of the site environment. In the latter case, the variations of χ_i are driven by the changes in the Madelung potential. For example, since in absolute value V_i is smaller on surface atoms than on bulk atoms, the cation electronegativity is expected to be higher and the oxygen electronegativity to be lower on surfaces than in the bulk. Similarly, the cation electronegativity will be higher on surfaces with higher Miller indices.

Equations (6.4.7) and (6.4.8) show that, in a Hartree scheme, there is a direct correspondence between χ_i and the position of the effective outer levels ϵ_i. We have seen in Chapters 1 and 3 that the effective atomic energies ϵ_i strongly determine the electron sharing between cations and oxygens in oxides, the oxygen charge being a decreasing function of the ratio $Z\beta^2/(\epsilon_M - \epsilon_O)^2$. The parameters which fix ϵ_M have just been discussed. The value of the cation ionic radius, on the other hand, is involved in the anion–cation first-neighbour distance R, upon which depend the resonance integrals β and, to a lesser extent, the Madelung potential values. On the other hand, ϵ_O does vary from one oxide to another, because it is a self-consistent function of the charge Q_O. At this point it is interesting to note that, in a given atom, parallel shifts of the outer and inner atomic levels are expected. This was shown, for example, by Pacchioni and Bagus (1994), who assigned the variations in the oxygen 1s binding energy in the alkaline-earth oxide series to the changes in the Madelung potential. The correlation between the oxide ionicity and the 1s core-level shifts pointed out by Vinek *et al.* (1977) may also be rationalized in that way, provided that one neglects the final state effects in the photoemission process.

While, in the bulk, the oxygen charge varies monotonically with the coordination number Z, the same is not true at the surface. For a given material, on non-polar surfaces, it is generally found (Chapter 3) that the surface charges are very close to the bulk ones, despite possibly large reduction of Z along some surface orientations. This comes from a reduction of the energy difference $\epsilon_M - \epsilon_O$, associated with the reduction of the Madelung potentials on surfaces. It will turn out in the analysis of surface OH groups that the positions of the atomic levels on surface atoms are a factor which is more important than the charge values.

Mixed oxides

On a catalyst surface, it may sometimes be important to enhance or reduce the acid–base strength, for example when an irreversible adsorption of the reactants takes place. The technique of doping the catalyst with small cations or anions such as Li^+, Ca^{2+}, Ni^{2+}, SO_4^{2-} ions, etc., or forming mixed oxides are then employed. When one calcines mixtures of co-precipitated hydroxides at high temperature, chemically mixed oxides are produced which involve an intimate mixing of both types of oxygen–cation bonds. But when mechanically mixed oxides are obtained by powder compression, the mixed bonds occur only at the grain boundaries. Clear and systematic trends in the reactivity and acidity of mixed oxides are not presently available.

Burggraf *et al.* (1982) and Chin and Hercules (1982) analyzed the activity and the nature of the reactive sites on γ-alumina surfaces covered with metallic overlayers. They showed that a diffusion of the metallic atoms into the matrix decreases the ability of the surface to be reduced, while this does not occur when the metal atoms remain at the surface and form a superficial oxide.

By measuring core-level shifts by x-ray photoemission, and stretching frequencies by infra-red absorption, and by carrying out catalytic tests, Noller *et al.* (1988) found that the acid strength of a mixed oxide is equal to some average of the strengths of its constituents. Shibata *et al.*'s (1973) and Tanabe's (1978) results prove, on the other hand, that the maximum acidity $H_{0,max}$ increases with the average electronegativity of the cations, and as the mean oxygen charge decreases. Gervasini *et al.* (1993) are in agreement with both criteria.

As far as chemically mixed oxides are concerned, Tanabe (1981) assigned the generation of Lewis or Brønsted acidity to an excess of positive or negative charge on the anion–cation bonds. He assumed that both kinds of cations maintain their coordination numbers when the mixed oxide is formed, and that all the oxygens have the coordination characteristic of the major component oxide. For example, in TiO_2–SiO_2, the oxygen coordination number is assumed to be equal to 4 or 2 depending upon whether TiO_2 is in excess or not. According to Pauling's partial charge model, a charge excess then occurs on the oxygen-minority cation bond. In TiO_2–SiO_2 with TiO_2 in excess, the 4+ silicon charge is shared between four bonds, while the 2− oxygen charge is distributed on three bonds. This yields a net positive charge equal to $+1 - 2/3 = 1/3$ on each Si–O bond. As a consequence Lewis acidity is generated. When SiO_2 is in excess, a similar calculation gives a sharing of the titanium 4+ charge between six bonds, a distribution of the oxygen −2 charge on two bonds and thus an excess negative charge equal to $4/6 - 1 = -1/3$ on each Ti–O bond. Around each titanium, a net negative charge equal to −2 is produced, which may bind two protons: Brønsted acidity is generated. If oxygens have the same coordination number in both oxides, as is the case in ZnO–ZrO_2 mixed oxide, no excess or loss of charge occurs and the acidity of the catalyst is not modified.

Seiyama (1978) proposed an alternative mechanism of acidity generation in mechanically mixed oxides. He assumed that the acidity appears at the boundary between the two oxides, and he calculated the oxygen charge as the sum of bond charges in each oxide. In ZnO–ZrO_2, for example, the oxygen has a negative charge equal to $-2 + 2/4 + 4/8 = -1$. In this estimation, −2 is the oxygen formal charge and the zinc and zirconium atoms respectively provide it with $+2/4$ and $+4/8$ electrons. A proton may come and neutralize this negative charge, which generates Brønsted acidity.

6.5 Hydroxyl groups on oxide surfaces

This last section focuses on the characteristics of hydroxyl groups on oxide surfaces, which have been the subject of detailed investigations in the context of acid–base interactions, but also whenever the surfaces are in contact with the ambient atmosphere, with water or with a biological medium. In toxicology, for example, this is the case for small silica particles which enter the lungs and are responsible for silicosis (Pézerat, 1989). It also occurs in colloid physics, in geology, electrochemistry, etc. In some cases, the interaction with water leads to disastrous effects, such as corrosion or adhesive or paint decohesion. But the water photo-dissociation at the surface of some oxides, such as TiO_2 or $SrTiO_3$, may be used to produce oxygen and hydrogen, which is a potential way of converting solar energy.

The first stages of adsorption, and more specifically the conditions under which water dissociates, turn out to be very important to understanding adhesion processes and more generally the surface reactivity, because the presence of protons and hydroxyl groups induces Brønsted acidity. The influence of such factors as temperature, surface coverage and defect concentration – oxygen vacancies, facets, etc., – on water dissociation has been considered carefully.

Experimental results

Thiel and Madey's review (1987) gives a good overview of the experimental studies on water dissociation at surfaces. Most of these studies have been performed on powders which become hydroxylated more easily than planar surfaces. This effect is assigned to the high density of defects on powders. It is confirmed by several studies of water dissociation on planar surfaces with low densities of steps (Brookes *et al.*, 1987; Onishi *et al.*, 1987; Sanders *et al.*, 1994).

On TiO_2 and $SrTiO_3$ surfaces, early studies reported water dissociation only for large concentrations of defects, such as Ti^{3+} species or oxygen vacancies. However, more recent experiments show that this occurs also on perfect {110} rutile surfaces up to a coverage of 0.1 monolayer, whatever the defect concentration (Kurtz *et al.*, 1989; Pan *et al.*, 1992). The analysis of band-bending by ultra-violet photoemission suggests that the surface bears a net negative charge. Oxygen vacancies on this surface were shown to bind preferentially to the hydroxyl groups (Hugenschmidt *et al.*, 1994), but on the (001) face, whether it is stoichiometric or (1×3) reconstructed, molecular water only was observed (Henderson, 1994). On the (100) face, the presence of steps or defects does not modify the surface reactivity versus water dissociation (Muryn *et al.*, 1991).

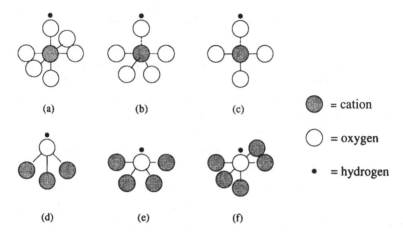

Fig. 6.5. Configurations of OH⁻ groups on oxide surfaces: (a), (b), (c) the hydroxyl group is adsorbed on a five-, four- and three-fold coordinated surface cation, respectively; (d), (e), (f) the proton is adsorbed on a three-, four- and five-fold coordinated surface oxygen, respectively.

Unlike rutile, the stoichiometric $SrTiO_3\{001\}$ faces do not dissociate water at room temperature. The same is true for the tungsten $Na_{0.7}WO_3$ and lead PbO oxides. However, on lead oxide, isolated oxygen atoms may be dissociation centres. On the (110) face of casserite SnO_2, the strongest dissociation of water molecules takes place for a low density of vacancies (Gercher and Cox, 1995).

On Al_2O_3 surfaces, a coexistence of several types of OH⁻ groups and water molecules has been found (Coustet and Jupille, 1994a; Frederick et al., 1991b). On the reconstructed (100) face of CuO_2, terminated with copper atoms, hydroxyl groups and water molecules are detected at 110 K, while only dissociative adsorption takes place at 300 K (Cox and Schulz, 1991). Thanks to adsorption isotherm measurements and infra-red absorption, hydroxyl groups have been identified on $BeO(10\bar{1}0)$ (Miyazaki et al., 1985). On thin nickel oxide films, dissociation of water may occur on the polar $\{111\}$ faces, but it depends upon the oxide thickness.

More recently, photoemission (Onishi et al., 1987) and HREELS (Shido et al., 1989; Wu et al., 1991a; Coustet and Jupille, 1994b) experiments have been performed on MgO. OH⁻ groups have been identified by photoemission on the $\{001\}$ face and on the faceted $\{111\}$ face, and their oxygen 1s levels have been shown to be shifted towards larger binding energies, by 2.5 eV with respect to the bulk.

Infra-red absorption and HREELS give access to the values of the OH⁻ stretching frequencies. When water, acids or bases are dissociated on a surface, several types of OH⁻ groups are formed. Some result from the adsorption of a proton on a surface oxygen. Others are pre-formed OH groups which adsorb on surface cations (Boehm, 1971). The stretching frequencies depend not only on the type of OH$^-$ group but also on the surface site and on the number of bonds formed between the OH$^-$ group and the surface (Fig. 6.5). There exist some correlations between the stretching frequency shift and the nature of the adsorption site (Coluccia *et al.*, 1988; Shido *et al.*, 1989; Morrow, 1990; Knözinger *et al.*, 1993a; 1993b):

- Higher frequencies are assigned to hydroxyl groups bound to surface cations, and lower frequencies to protons adsorbed on surface oxygens. Both families present many internal splittings on rough surfaces, due to all the possible surface sites, adsorption sites – on-bridge, ternary, etc., – and lateral interactions between adsorbed groups.
- On a given compound, lower stretching frequencies characterize less basic oxygens because they reveal weaker O–H bonds. For hydroxyl groups adsorbed on a surface cation, lower frequencies characterize more acid cations, because a weaker O–H bond is generally associated with a stronger OH–cation bond. This rule is not always obeyed when different compounds are compared.
- The stretching frequency of hydroxyl groups bound to a single surface cation is lower on more under-coordinated cations. The reverse is true for protons adsorbed on-top a surface oxygen.

Information on the preferential bonding between adsorbates and surface ions, and on water dissociation may also be obtained by secondary ion mass spectrometry, in the limit of low surface damage. When hydrogen or oxygen isotopes are used to mark water (e.g. D_2O or $H_2^{18}O$), the surface oxygens may, in addition, be distinguished from those belonging to the water molecules in the fragments (Bourgeois *et al.*, 1992).

Recently, *local* acid–base properties have been determined, on a length scale of about 400 Å, by measuring the double layer forces, between a surface and the tip of an atomic force microscope immersed in a liquid (Lin and Arribart, 1993; Lin *et al.*, 1993).

Theoretical results

Most numerical approaches have considered the adsorption of protons and hydroxyl groups on surfaces in the limit of zero coverage. By modelling the surfaces by small clusters of a few atoms, on the basis of static energetic considerations, they have studied whether a dissociative adsorption occurs, and have calculated the charge transfers between H^+ or OH^- and the surface and OH stretching frequencies. They have focused on the (110) faces of TiO_2 and the (001) face of $SrTiO_3$ (Tsukada *et al.*, 1983), several faces of TiO$_2$ (Fahmi and Minot, 1994), silica surfaces (Ugliengo *et al.*, 1990), active sites on zeolites (Brand *et al.*, 1992), and MgO(100) (Abarenkov *et al.*, 1987; Langel and Parrinello, 1994). Pak (1974) reported trends followed by stretching frequencies and charge transfers as a function of the oxide electronegativity. In a similar way, Pelmenschikov *et al.* (1991) considered OH$^-$ adsorbed on alumina, silica, phosphorus and boron oxides. The adsorption energies and the electronic characteristics of protons and hydroxyl groups, adsorbed on the following oxides: BaO, SrO, CaO, MgO, TiO$_2$ SiO$_2$ and SrTiO$_3$ and on various faces of MgO, in the limits of zero coverage and full saturation of the surface have also been calculated (Russo and Noguera, 1992a; 1992b; Goniakowski *et al.*, 1993; Goniakowski and Noguera, 1993; 1994c; 1995b; Noguera *et al.*, 1993). To the author's knowledge, the only other attempts to treat fully-hydroxylated surfaces were performed on ZrO$_2$ (100) (Orlando *et al.*, 1992) and MgO (100) (Scamehorn *et al.*, 1993; 1994). In these treatments, a slab geometry was used rather than an embedded cluster geometry.

Due to the development of advanced numerical methods in the last decades, quantum approaches are now able to accurately describe the chemical bonds formed between two reactants. Nevertheless, when a surface is involved, the actual systems met in practice, for example a dense polymeric layer adsorbed on a rough surface, cannot yet be simulated, because this would require too large a memory size or too long a computation time. Quantum calculations, thus, cannot compete with empirical models in the prediction of adhesion strengths. However, they may allow one to check their validity in model cases, for example small molecules adsorbed on a substrate, or large molecules adsorbed on a cluster of a few atoms which simulates the substrate. This has been done in a number of cases but, to the author's knowledge, mostly for adsorption processes on metallic surfaces. Numerical results for the adsorption of molecules on oxide surfaces may be found in the literature (Henrich and Cox, 1994), but there exists no systematic discussion in the framework of acid–base interactions.

In order to point out which factors drive the acid strengths, it is helpful to discuss in detail the results obtained from a quantum calculation of the adsorption of protons and hydroxyl groups. The adsorptions of protons and hydroxyl groups represent elementary acid–base reactions involving a surface. The calculations yield several quantities which are relevant in the context of acid–base interactions: the charge transfers along the interfacial bonds; the adsorption energies; and the geometric characteristics of the OH groups when an optimization procedure is used. The first quantity is directly related to the Lewis acidity or basicity, i.e. to the ability to give or receive electrons. The second is of interest to discuss, e.g., the *IEPS* value, which is related to the free energies of protonation reactions, or the Brønsted acidity. By considering various oxides, various surface orientations and several adsorbate densities, it is possible to discuss the acid–base strength as a function of the oxide ionicity, as a function of the nature of the surface sites and as a function of the surface coverage.

Two series of oxides will be considered: the series BaO(100), SrO(100), CaO(100), MgO(100), TiO$_2$(110), and SiO$_2$(0001) (Goniakowski *et al.*, 1993); and three surface orientations of MgO: (100), (110) and (211) (Goniakowski and Noguera, 1994c; 1995b), both in the limit of zero coverage and at full saturation of the surface by dissociated water molecules. This will allow one to discuss the parameters which determine the acid–base strength of oxide surfaces and the applicability of empirical approaches to these adsorption processes in situations of increasing complexity: 1) a series of oxides presenting different ionicities; 2) several surface orientations of a given oxide; and 3) two coverages of a given oxide surface.

From a geometric point of view, the protons always adsorb on-top of a surface oxygen, because the small proton ionic radius prevents the formation of two bonds with surface atoms. In the first series of oxides, the hydroxyl groups adsorb on a single surface cation with their oxygen at the position of a missing lattice oxygen. The adsorbate coordination number Z_a is thus equal to unity ($Z_a = 1$). In the second series, they may bind to several cations. On MgO(100) the most likely adsorption configuration is on-top ($Z_a = 1$), while, on the (110) face, the OH$^-$ groups may adsorb on-top ($Z_a = 1$) or on-bridge ($Z_a = 2$), and on the (211) face they may adsorb on-top ($Z_a = 1$), on-bridge ($Z_a = 2$) or in a ternary site ($Z_a = 3$). Usually, the adsorption site with the largest Z_a is the most stable. We will focus successively on the charge transfers, the densities of states, the adsorption energies and the structural characteristics of OH groups, both in the limit of zero coverage and in the limit of full saturation of the surface by dissociated water molecules.

Charge transfers

When a proton adsorbs on a surface oxygen, an electron transfer n_{H^+} takes place from the surface towards the proton. The sign of the transfer is reversed between an OH$^-$ group and a surface cation ($\Delta n_{OH^-} < 0$). In the latter case, there is also a modification of the O–H bond, the terminal hydrogen bearing less electrons when the transfer to the surface is larger. By its very definition, the importance of the adsorbate–substrate charge transfer is a measure of the Lewis acidity or basicity.

Table 6.1 shows that n_{H^+} decreases as the oxide ionicity decreases, while Δn_{OH^-} increases in absolute value and reaches especially high values on the rutile and quartz surfaces. In the limit of full saturation of the surface by

Table 6.1. *Charge transfers for protons and hydroxyl groups adsorbed on various oxide surfaces, in the limit of zero coverage* (top part of the table) *and in the limit of saturation* (bottom part).

	BaO (100)	SrO (100)	CaO (100)	MgO (100)	TiO$_2$ (110)	SiO$_2$ (0001)
n_{H^+}	0.83	0.81	0.78	0.76	0.65	0.60
Δn_{OH^-}	−0.07	−0.08	−0.11	−0.13	−0.54	−0.88
n_{H^+}	0.67	0.65	0.61	0.59	0.63	0.63
Δn_{OH^-}	−0.31	−0.33	−0.36	−0.40	−0.47	−0.75

*n_{H^+} is the electron number on the proton and Δn_{OH^-} the change in the hydroxyl group electron number.

dissociated water molecules, there is a modification of the charge transfers. The interaction between adsorbates induces an important weakening of the surface–proton charge transfer on rocksalt surfaces (about 0.15 electron), but nearly no variation on TiO$_2$ and SiO$_2$. An effect of opposite sign is found for Δn_{OH^-}, which becomes much larger on rocksalt surfaces and decreases slightly on rutile and quartz. However, the change of n_{H^+} and Δn_{OH^-} along the series remains qualitatively unchanged, whatever the density of adsorbates.

When the surfaces are hydroxylated, a surface dipolar field is present which involves the hydration layer and the few first atomic substrate layers. The dipole sign is determined by a competition between the two types of transfers discussed above. The hydroxylation layer is negatively charged on the rocksalt surfaces and TiO$_2$ and positively charged on SiO$_2$. The result on TiO$_2$(110) is in agreement with the sign of the band-bending measured by Kurzt *et al.* (1989).

Table 6.2 shows the corresponding charge transfers for the three MgO surfaces. While n_{H^+} remains roughly constant in the series, Δn_{OH^-} grows as the surface indices become larger. Yet, since the number of bonds Z_a (a = OH$^-$) between the adsorbate and the surface increases linearly in the series (Z_{OH^-}=1, 2 and 3, respectively), the charge transfer *per bond*: $\Delta n_{OH^-}/Z_{OH^-}$, is nearly independent of the surface orientation. The interaction between adsorbates yields large modifications of the charge transfers on MgO(100). n_{H^+} decreases and Δn_{OH^-} increases. The effect vanishes as the density of the hydroxylation layer goes to zero.

As far as electron transfers are concerned, i.e. in the framework of Lewis acidity or basicity, it may, thus, be concluded that the oxides from BaO

Table 6.2. *Charge transfers
for proton and hydroxyl-group
adsorption on three MgO surfaces.*

MgO	(100)	(110)	(211)
Z_{OH^-}	1	2	3
n_{H^+}	0.76	0.76	0.77
Δn_{OH^-}	−0.13	−0.26	−0.34
n_{H^+}	0.59	0.66	0.74
Δn_{OH^-}	−0.40	−0.33	−0.33

*Same notations as Table 6.1.

to SiO_2 become more and more acid, both because their surface oxygens donate less and less electrons and because their surface cations accept more and more electrons. This change may be correlated with the increasing covalency of the cation–oxygen bond in the series. The conclusion does not depend upon the adsorbate density, which only changes the strength of the adsorbate–substrate acid–base interaction. The basicity of rocksalt oxides and the acidity of rutile and quartz decrease when the coverage of the surface increases.

On the three MgO faces, the charge transfer *per bond* does not depend much upon the surface orientation in the limit of zero coverage. When the density of adsorbates becomes large, the conclusion is modified. It is found that surfaces become more and more basic as the coordination number of the surface atoms decreases.

These results may be rationalized using the theoretical tight-binding model developed in Chapters 1 and 3, which takes into account the local hybridization around each atom. The electron transfers are increasing functions of the ratio $Z_a \beta^2/(\epsilon_1-\epsilon_2)^2$, where Z_a is the number of adsorbate–substrate bonds, β the effective resonance integral between the two species, and $\epsilon_1 - \epsilon_2$ the energy difference between the donor and acceptor energy levels. The parameters which control these quantities will be sequentially discussed: the adsorbate–substrate bond length; the number of adsorbate–substrate bonds; and the value of $\epsilon_1 - \epsilon_2$.

Adsorbate–substrate bond length   The value of the adsorbate–substrate bond length controls the strength of the resonance integrals β. In the case of proton adsorption, this factor cannot discriminate between the various oxides, because the proton–surface bond length remains very close to the O–H inter-atomic distance in the free hydroxyl group or in the water molecule. In the case of hydroxyl group adsorption, the oxygen atom belonging to the OH^- group roughly adsorbs at the position of a missing lattice oxygen. The O–cation distance is, thus, a linear function of the cation ionic radius. The relevance of the cation ionic radius, which has been stressed in Section 6.4, thus, appears in quantum approaches through the values of the resonance integrals. For a given type of orbital, resonance integrals are larger for smaller inter-atomic distances, even when the principal quantum number changes (Harrison, 1981; Hoffmann, 1963). Such a conclusion does not apply when comparing resonance integrals which involve different types of overlaps.

The value of the Madelung potential exerted by one species on the other, which renormalizes the effective atomic level positions ϵ_1 and ϵ_2, also depends on the adsorbate–substrate distance.

Number of adsorbate–substrate bonds   As already mentioned, a proton cannot form multiple bonds with a planar oxide surface, because its radius is too small to allow a close approach to two surface oxygens. On the other hand, the oxygen of an hydroxyl group may bind to several cations. The Z_a value, thus, depends upon the surface orientation and upon the presence of surface irregularities. According to Table 6.2, the charge transfer Δn_{OH^-} is larger for higher Z_a, in the limit of zero coverage, which is in agreement with the increase of the ratio $Z_a \beta^2 / (\epsilon_1 - \epsilon_2)^2$.

Energy difference between the adsorbate and substrate atomic energy levels   The presence of $\epsilon_1 - \epsilon_2$, the energy difference between the acceptor and donor renormalized atomic levels, in the denominator of the ratio $Z_a \beta^2 / (\epsilon_1 - \epsilon_2)^2$, implies that the charge transfer increases when $\epsilon_1 - \epsilon_2$ becomes smaller. This extends earlier approaches, which treated the resonance integrals by perturbation and assumed that the atomic energies were those of the neutral atoms (Mulliken, 1951; 1952a; 1952b; Hudson and Klopman, 1967; Klopman and Hudson, 1967). For proton adsorption, ϵ_1 is equal to the hydrogen 1s atomic energy, and ϵ_2 to the surface oxygen 2p atomic energy. For OH^- adsorption, ϵ_1 is the energy of the surface cation outer level and ϵ_2 the energy of the OH^- oxygen 2p level (Fig. 6.6).

Since all these atomic energies have charge and Madelung corrections in a Hartree scheme, the energy difference $\epsilon_1 - \epsilon_2$ depends upon the cation electronegativity, the surface oxygen charge, the oxide structure and the surface coordination number. As the cation electronegativity increases in the first oxide series, ϵ_1 shifts to lower energies and gets closer to the energy ϵ_2 of the OH^- group. The charge transfer Δn_{OH^-} increases. Pak (1974) found a similar result, along the series: $(-O)_5 MgOH$, $(-O)_5 TiOH$, $(-O)_5 AlOH$, $(-O)_3 SiOH$.

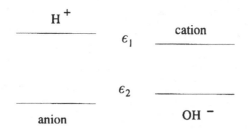

Fig. 6.6. Atomic levels involved in the formation of a chemical bond between a proton and a surface anion (*on the left*) or an hydroxyl group and a surface cation (*on the right*). In the latter case, the filled bonding level of the OH^- group is represented.

Along the oxide series, the surface oxygen level ϵ_2 shifts towards lower energies as a result of the reduction of the oxygen charge $|Q_O|$. The energy difference $\epsilon_2 - \epsilon_1$ involved in the proton adsorption, thus, increases, yielding smaller and smaller charge transfers n_{H^+}.

The Madelung corrections to the atomic energy level position have to be carefully taken into account when several surfaces of a given compound are considered, as in the case of the MgO series. We have already noted that the Madelung potential shifts the surface oxygen levels towards higher and higher energies and the surface cation levels towards lower and lower energies, in this series. In addition, it also shifts the adsorbate levels with an increasing strength, because of the reduced atomic density in the outer plane (inter-plane interactions increase, as intra-plane interactions decrease). It raises the proton level and lowers the hydroxyl group levels in a more and more efficient way in the series. The energy difference $\epsilon_1 - \epsilon_2$ happens to be roughly constant, both for proton and hydroxyl group adsorption. This explains why n_{H^+} remains constant on the three surfaces and why the electron transfer *per bond* $\Delta n_{OH^-}/Z_a$ is roughly constant for the adsorption of hydroxyl groups.

The adsorbates also exert an electrostatic potential on the substrate atoms and on neighbouring adsorbates. The effect is especially noticeable when the adsorbate density is large. Fig. 6.7 exemplifies the energy level shifts induced by the Madelung potential of the hydroxylation layer on MgO(100). On the surface–proton bond, the electrostatic potential exerted by the neighbouring OH^- ions is larger on the protons than on the surface oxygens. $\epsilon_1 - \epsilon_2$ is, thus, larger than in the case of the adsorption of a single molecule. The charge transfer decreases. On the cation–OH bond, the electrostatic potential exerted by the neighbouring protons is roughly equal on the substrate and on the adsorbate. The second-neighbour OH^- groups, on the other hand, exert a larger potential on the hydroxyl group than on the surface cation. This decreases $\epsilon_1 - \epsilon_2$ and induces a larger

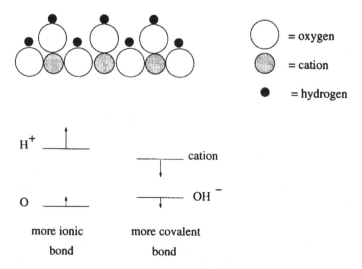

Fig. 6.7. Representation of a saturated hydroxylation layer on a (100) face of a rocksalt oxide. In the lower part of the figure, the shifts of the adsorbate and substrate atomic energy levels, induced by the presence of the neighbouring adsorbates, are shown by arrows.

charge transfer than in the case of the adsorption of a single molecule. On this surface, the presence of the hydroxylation layer, thus, induces a lower oxygen basicity and a larger cation acidity. However, the above conclusion crucially depends upon the surface geometry and has to be reconsidered for each particular case. For example, the charge transfer between OH^- and $SiO_2(0001)$ is lower on a fully-hydroxylated surface than for the adsorption of a single molecule, contrary to what happens on the (100) rocksalt oxide surfaces.

To summarize, we have stressed that the Lewis acidity depends upon three parameters which are: the adsorbate coordination number Z_a, i.e. the number of interfacial bonds per adsorbed molecule; the resonance integral β; and the acceptor–donor energy difference $\epsilon_1 - \epsilon_2$. These quantities are related to the cation electronegativity, the surface oxygen charge, the oxide structure, the surface orientation and the surface coverage, but the relationship is generally intricate due to the self-consistent link between charges and potentials, and due to the mutual interaction between the adsorbate and the substrate.

Density of states

The formation of interfacial bonds modifies the local densities of states (LDOS: local density of states) on the substrate atoms involved in the bond. Generally, the adsorption of a proton on a surface oxygen is

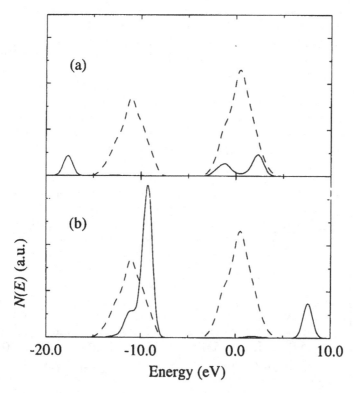

Fig. 6.8. Local density of states for the adsorption of: (a) a proton and (b) an hydroxyl group on MgO(100) in the limit of zero coverage (arbitrary units). Full and dashed lines represent the LDOS on the hydrogen atom and on surface atoms, respectively.

associated with a strong bond, reminiscent of the O–H bond in the free hydroxyl group. The proton–surface oxygen inter-atomic distance is indeed close to the value 0.957 Å found in the free hydroxyl group. The signature of the bonding and anti-bonding states of the newly formed OH entity, thus, appears in the surface oxygen LDOS (Fig. 6.8a). The bonding state has an energy close to the valence band states and the anti-bonding state is generally located above the conduction band. The exact positions vary in the different oxides.

The cation–OH⁻ bond formation is very dependent upon the parameters which have been discussed above: cation ionic radius; adsorption site; and Madelung potential (Fig. 6.8b). When the bond is weak, as on the rocksalt oxide surfaces, the LDOS on a surface cation is little perturbed and the states of the OH⁻ molecule are superimposed on the oxide band structure without noticeable interferences. When the bond is strong, which happens on more acid oxides such as rutile and quartz, a quasi-SiOH or

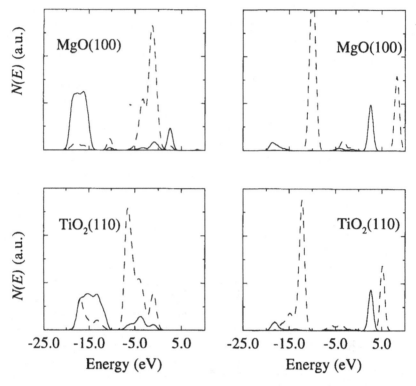

Fig. 6.9.   Local densities of states (LDOS) on hydroxylated MgO(100) and TiO$_2$ (110) surfaces (arbitrary units). *On the left*: oxygen (full lines) and cation (dashed lines) LDOS; *on the right*: adsorbed proton (full lines) and hydroxyl group (dashed lines) LDOS.

-TiOH molecule forms, whose bonding, anti-bonding and non-bonding states have non-negligible weights on the three atoms.

On fully-hydroxylated surfaces, the local density of states is more strongly perturbed, because all the atoms of the substrate top layer take part in the interfacial bonds. The proton and hydroxyl group levels are spread into bands. The surface anion and cation levels are lowered and raised, respectively, due to the electrostatic potential exerted by the hydroxylation layer, so that the gap width in the LDOS of these atoms is larger than on the free surface. In most cases, it is even wider than in the bulk. Bulk states, thus, constitute the bottom of the substrate conduction band. Depending upon the compounds and the surfaces under consideration, the filled states of the adsorbed OH$^-$, or more precisely the bonding states of the substrate–OH$^-$ molecule, may be located at various positions relative to the bottom of the conduction band. It is this energy difference which determines the gap width of the hydroxylated surface. The densities of states of some hydroxylated oxide surfaces are represented in Fig. 6.9.

Adsorption energies

While the Lewis acidity and basicity strengths are related to the ability to give or receive electrons and may be quantified by the value of the adsorbate–substrate charge transfers, some experimental quantities, also related to the surface acid–base character, are functions of the *adsorption energy* of the species. This is true for the isoelectric point $IEPS$ whose value is determined by the free energies of protonation reactions. It is also true for the stretching frequencies which are related to the curvature of the bond energy curve around the equilibrium distance. The adsorption energy involves several terms, among which the electrostatic (Madelung) energy and the covalent energy are the most relevant (see Chapter 1). Since these two contributions vary with the charge transfers neither in a proportional nor in a monotonic way, it may be expected that the oxide acidity scale depends upon whether one uses a charge-transfer criterion (Lewis' acidity) or an adsorption energy criterion ($IEPS$, adsorption of test molecules, etc.). Tables 6.3 and 6.4 give the values of the adsorption energies of protons and hydroxyl groups in the series of oxides and for the three MgO faces, respectively.

Along the series of oxides, in the limit of zero coverage, the proton adsorption energy $E_{\mathrm{H}^+}^{\mathrm{ads}}$ decreases while the hydroxyl group adsorption energy $E_{\mathrm{OH}^-}^{\mathrm{ads}}$ increases. This trend is consistent with the values of the

Table 6.3. *Adsorption energies (in eV) of protons and hydroxyl groups on various oxide surfaces.*

(100)	BaO (100)	SrO (100)	CaO (100)	MgO (110)	TiO$_2$ (0001)	SiO$_2$
$E_{\mathrm{H}^+}^{\mathrm{ads}}$	10.7	10.1	9.3	8.3	5.9	4.8
$E_{\mathrm{OH}^-}^{\mathrm{ads}}$	0.4	0.0	0.3	1.0	8.3	10.3
$E_{\mathrm{tot}}^{\mathrm{ads}}$	11.1	10.1	9.61	9.3	14.2	15.1
E^{ads}	12.7	12.2	12.1	12.3	16.6	21.0

$^\ast E_{\mathrm{H}^+}^{\mathrm{ads}}$ and $E_{\mathrm{OH}^-}^{\mathrm{ads}}$, respectively, are the proton and hydroxyl-group adsorption energies in the limit of zero coverage. $E_{\mathrm{tot}}^{\mathrm{ads}}$ is their sum and E^{ads} the adsorption energy per dissociated water molecule in the limit of saturation of the surface.

charge transfers discussed above. A non-monotonic variation of E_{tot}^{ads} results ($E_{tot}^{ads} = E_{H^+}^{ads} + E_{OH^-}^{ads}$). E_{tot}^{ads} decreases along the rocksalt series but becomes large again for rutile and quartz. In the saturation limit, E^{ads} is roughly constant in the rocksalt series and becomes larger for rutile and quartz.

On the three MgO surfaces, the proton and hydroxyl-group adsorption energies increase, and so do E_{tot}^{ads} and E^{ads}. One must note that, in both series, the interactions between adsorbates generally favour the adsorption. The dissociative adsorption energy of water molecules – equal to ($E_{tot}^{ads} - E_{H_2O}^{diss}$) or ($E^{ads} - E_{H_2O}^{diss}$), depending upon the coverage – is negative for the (100) faces of rocksalt oxides and positive for the two faces of TiO_2 and SiO_2. In the last two cases, the surfaces will get hydroxylated in the presence of a humid atmosphere, if the conditions to overcome the activation barrier to dissociation are fulfilled. The absence of dissociative adsorption on a perfect MgO(100) surface is due to the weak value of $E_{OH^-}^{ads}$. However, this restriction no longer exists on MgO(110) and (211) faces, on which the OH$^-$ groups may bind to two and three surface cations, respectively.

To summarize, as far as the adsorption energies are concerned, it may be said that an increased covalency of the oxides is associated with a larger acidity. The surface oxygens are less basic and the surface cations more acidic. The results obtained on the three MgO faces, on the other hand, agree with the general statement that more under-coordinated sites are generally more reactive. We find that the oxygens are more basic and the magnesiums more acidic in the series.

Table 6.4. *Adsorption energies (in* eV*) of protons and hydroxyl groups on three* MgO *surfaces.*

MgO	(100)	(110)	(211)
$E_{H^+}^{ads}$	8.3	9.1	11.6
$E_{OH^-}^{ads}$	1.0	2.4	4.9
E_{tot}^{ads}	9.3	11.5	16.5
E^{ads}	12.3	15.8	19.7

*Same notations as Table 6.3.

At variance with the electrostatic models, the covalent contribution to the adsorption energy generally prevails in the strong chemisorption processes discussed here. This means that they are driven by the formation of covalent interfacial bonds between the adsorbates and the substrate. A generalization of Equation (1.4.62), similar to that made for the case of charge transfers, allows one to write the covalent contribution to the bond energy in the following way :

$$E_{\text{cov}} = \frac{-4n_0 Z_a \beta^2}{\sqrt{(\epsilon_1 - \epsilon_2)^2 + 4Z_a \beta^2}} . \tag{6.5.1}$$

It is an increasing function of the ratio $Z_a \beta^2/(\epsilon_1 - \epsilon_2)$. (Note that, at variance with the charge transfer expression, the energy difference in the denominator is not squared). In the limiting case where β is much smaller than $\epsilon_1 - \epsilon_2$, this gives the expressions found in the works of Mulliken (1951; 1952a; 1952b) and Hudson and Klopman (1967). The parameters on which the ratio $Z_a \beta^2/(\epsilon_1 - \epsilon_2)$ relies have been discussed above. As far as the surface–OH^- bond is concerned, the covalent energy is weak on basic oxides and strongly increases on acidic oxides. This behaviour is consistent with the values of the charge transfers which have been discussed above and can be explained with the same arguments – increasing electronegativity, decreasing ionic radius, increasing Madelung potentials. Similarly, on the three MgO surfaces, despite the fact that the charge transfer on each Mg–OH bond is roughly constant, the formation of 1, 2 and 3 bonds in the series induces an increase of the *total* charge transfer and of the covalent energy. A simple model of chemisorption, in the limit of weak coupling, displays the same trends for metallic surfaces (Desjonquères and Spanjaard, 1982; 1983).

The electrostatic energy, on the other hand, is not simply equal to the direct charge–charge interaction between the adsorbate and the surface atom on which it adsorbs, as assumed in Parks' or MUSIC models, because of the long range of the coulomb interactions and because of the presence of the adsorbate–substrate charge transfers. Several points are worth noticing:

- The adsorbate–substrate interaction is the result of attractive and repulsive interactions with all the substrate ions. It is thus smaller than the direct interaction with a single ion. For example, on MgO(100), assuming integer charge values, the proton–oxygen interaction is of the order of 30 eV, i.e. about fifteen times larger than the proton–surface interaction. It should be estimated not with the substrate ionic formal charges but with the effective charges resulting from oxygen–cation electron sharing.

- The adsorbate–substrate charge transfer is responsible for a charge decrease of both species. The hydroxyl groups and surface oxygens lose electrons while protons and surface cations capture electrons. This also reduces the direct adsorbate–substrate interaction, and the effect is more and more effective as the covalency of the interfacial bond becomes larger.
- The charge decrease on the adsorption site also modifies the electrostatic interactions *inside* the substrate. The cohesion of the latter decreases. For a given charge transfer, the effect is larger on dense surface planes. Actually, this electrostatic decohesion explains the increase of the proton adsorption energy in the series of the three MgO surfaces. This is a typical case in which one would conclude that, on the basis of charge transfer considerations, the oxygen basicity does not depend upon the surface orientation, while adsorption energy considerations lead to the conclusion of an increasing basicity in the series.

This discussion also shows that neither the covalent nor the electrostatic energies can be written as simple products of parameters characterizing the acid A and the base B, as postulated in the Drago and Wayland (1965) $\mathscr{E}\&\mathscr{C}$ equation, (6.1.15). The properties of the coupled system – for example, the number of formed bonds between the adsorbate and the substrate or the substrate decohesion which results from the charge transfer – are also important. The same remarks apply to the model of adhesion proposed by Bolger (1983) which expresses the adhesion energy as the difference between two energies characteristic of the reactants in the absence of each others.

Structural characteristics of hydroxyl groups

Table 6.5 gives the inter-atomic O–H distances obtained for proton and hydroxyl group adsorption on the three MgO faces. The quantity d_{OH^+} denotes the interfacial O–proton bond length and d_{OH} the O–H inter-atomic distance inside the hydroxyl group. The values of the proton n_{H^+} and of the OH terminal hydrogen n_H electron numbers are also written. One should note that the values of n_{H^+} obtained after the geometry optimization, which are given in Table 6.5, are slightly different from those obtained when assuming a rigid geometry (Table 6.2).

As a general trend, in the limit of zero coverage, the O–H and OH$^+$ bond lengths are very close to that of the free hydoxyl group ($d_{OH} = 0.957$ Å). They do not vary significantly with the surface orientation. On the other hand, on a fully-hydroxylated surface, both d_{OH^+} and d_{OH} decrease in the series. A clear correlation between the O–H bond lengths and the hydrogen

Table 6.5. *Inter-atomic O–H distances (in* Å*) for proton and hydroxyl-group adsorption on three MgO surfaces, in the limit of zero coverage* (top part of the table) *and in the limit of saturation* (bottom part).

MgO	(100)	(110)	(211)	MgO	(100)	(110)	(211)
d_{OH^+}	0.963	0.961	0.958	n_{H^+}	0.77	0.77	0.76
d_{OH}	0.960	0.960	0.964	n_H	0.85	0.80	0.76
d_{OH^+}	0.991	0.972	0.964	n_{H^+}	0.62	0.70	0.74
d_{OH}	0.982	0.966	0.964	n_H	0.66	0.73	0.75

*d_{OH^+} and d_{OH}, respectively, denote the interfacial oxygen–proton bond length and the O–H interatomic distance inside the hydroxyl group. The corresponding values of the proton n_{H^+} and the OH terminal hydrogen n_H electron numbers are also given.

electron numbers is revealed by Table 6.5. A bond length expansion takes place whenever the oxygen–hydrogen electron transfer decreases. This stresses the relationship between the surface acidity and the structural characteristics of the O–H bond: the equilibrium inter-atomic distance; and the stretching frequency.

As a general statement, it is recognized that, when a chemical bond forms, the dependence of the bond energy upon the bond length presents systematic features. For example, the deeper the well in the energy curve, the larger the curvature. A universal model, accounting for these features, was proposed for metallic cohesion, adsorption and adhesion processes (Rose *et al.*, 1983) and extended to the particular case of transition metals (Spanjaard and Desjonquères, 1984). It relies upon the assumption that the adsorption energy results from a competition between an attractive interaction and a short-range repulsion term, the latter varying with the bond length more quickly than the former (e.g. $n > m$ in (6.5.2) below).

In the context of acid–base interactions, it has been noted that systematic relationships between the bond lengths, the charge transfers and the coordination numbers exist and qualitative rules have been established (Lindqvist, 1963; Gutmann, 1978). They state that in the adduct produced by a donor–acceptor reaction, an expansion of bond lengths takes place inside the donor (D) and acceptor (A) parts of the adduct around the newly formed A–D bond, and that the expansion is larger when the A–D bond is shorter (first bond-length variation rule). It may also be expressed in terms of the increase in coordination number: as the coordination number increases, so do the lengths of the bonds originating from the coordination centre (third bond-length variation rule). The relaxations of distances are not confined to the bonds adjacent to the A–D bond. Intra-molecular changes are such that a σ-bond is lengthened when an electron transfer takes place from the more electropositive to the more electronegative atom in the uncomplexed species, and it is shortened otherwise (second bond-length variation rule).

It is possible to develop a theoretical model for the O–H bond, which includes these characteristics. It relies upon the assumption that, as a first approximation, the covalent energy is the attractive term whose distance dependence is the strongest. It is proportional to $\sqrt{n_H(2 - n_H)}$, n_H being the electron number borne by the hydrogen atom, according to the hetero-polar diatomic molecule model (Chapter 1, Equation (1.3.16)). No distinction between n_H and n_{H^+} will be kept in the following. Assuming, in addition, that the covalent energy varies as an inverse power law $(1/d^m)$ with the O–H distance d, and that the short-range repulsion term is written in a Lennard–Jones form (A/d^n), the d-dependent part of the O–H bond energy, thus, reads:

$$E = -\frac{B\sqrt{n_H(2 - n_H)}}{d^m} + \frac{A}{d^n} . \qquad (6.5.2)$$

The bond energy E_0, the equilibrium inter-atomic distance d_0 and the stretching frequency – which is proportional to the square root of $\partial^2 E/\partial d^2$ at $d = d_0$ – are, respectively, proportional to:

$$E_0 \propto [n_H(2 - n_H)]^{\frac{n}{2(n-m)}} , \qquad (6.5.3)$$

$$d_0 \propto \left[\frac{1}{n_H(2 - n_H)} \right]^{\frac{1}{2(n-m)}} , \qquad (6.5.4)$$

$$\nu_0 \propto [n_H(2 - n_H)]^{\frac{n+2}{4(n-m)}}. \qquad (6.5.5)$$

Their variations as a function of n_H are represented in Fig. 6.10, after normalization to the free hydroxyl group characteristics (for which $n_H = 0.97$ in this approach). Short bond lengths and high stretching frequencies are associated with strong bonds, characterized by large covalent charge transfers n_H. In the specific case of OH species, this model thus provides an analytical basis to the bond-length variation rules given above.

Equations (6.5.3)–(6.5.5) make the link between the oxygen Lewis basicity – i.e. the ability of the oxygen to give electrons to the hydrogen atom, measured by n_H – and the O–H structural characteristics: the interatomic distance; and the stretching frequency. High stretching frequencies are,

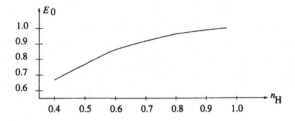

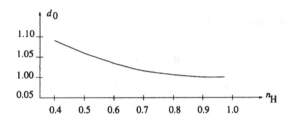

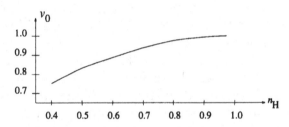

Fig. 6.10. Bond energy E_0, equilibrium inter-atomic distance d_0 and stretching frequency v_0 of an hydroxyl group as a function of the electron number n_H borne by the hydrogen atom, according to (6.5.3)–(6.5.5). Each quantity is normalized to its value in the free hydroxyl group ($n_H = 0.97$ in this model).

thus, expected for protons adsorbed on highly basic surface oxygens or for OH groups weakly interacting with the surface cations. The latter conclusion comes from the observation that a strong bond between the substrate and the OH group is always associated with a weak O–H bond. Conversely, on surfaces with strongly acidic cations, hydroxyl groups with high stretching frequencies are found. The fact that, on MgO(100), the proton–surface bond is weaker than the O–H bond in the adsorbed group, implies that the OH groups adsorbed on cations should have the highest stretching frequency. This is in agreement with the present understanding of experimental results (Shido *et al.*, 1989; Coustet and Jupille, 1994b). The relationship between the bond length and the charge transfer is similar in spirit to that noticed in inorganic crystals between the inter-atomic distances and the bond valences (Brown, 1977).

The structural characteristics of OH groups depend upon the density of adsorbates. In particular, on the fully-hydroxylated MgO(100) surface, the interaction between adsorbates induces a decrease of the O–H$^+$ charge transfer and, consequently, an expansion of the interfacial bond length. According to (6.5.3)–(6.5.5), a shift of the stretching frequency towards lower values should result. On less dense surfaces, the structural modifications of the OH groups when the density of adsorbates varies are weaker. The dependence of the structural properties of OH groups upon the adsorbate density is generally not well understood in the literature.

Synthesis

The understanding of the electronic and structural characteristics of OH$^-$ groups on oxide surfaces is more advanced than that of other molecules. Yet, it faces several difficulties. On the experimental side, the microscopic characterization of the adsorption sites is not easy because, in most studies, oxide powders are used. The model example of the MgO surfaces shows that, on a planar open surface such as (211), at least four non-equivalent OH$^-$ groups exist, if one takes into account protons and hydroxyl groups with coordination numbers equal to 1, 2 and 3. The analysis of planar monocrystalline surfaces, with a controlled amount of defects, should help in getting a better fundamental understanding of these processes.

From a detailed theoretical analysis of charge transfers, adsorption energies and structural properties of OH groups on various oxide surfaces, various orientations and coverages, several conclusions may be drawn. First, it turns out that the Lewis' acidity depends upon quantities which are related, in an intricate way, to the cation electronegativity, the surface oxygen charge, the oxide structure, the surface orientation and the surface coverage. The trends found for adsorption energies, on the other hand, do not systematically coincide with those found for charge transfers, thus revealing a sensitivity of the acidity scales to the parameter on which they are based. It is necessary to consider the oxygen and cation electronegativities, not only in their proper charge states in the oxide, but also modified by the Madelung potential exerted by all the neighbouring charges. This introduces a concept of electronegativity *in the solid*. The analytical formulation of the ionicity of the oxygen–cation bond in oxides and of the adsorbate charge transfer is available, which extends earlier approaches based on empirical arguments or perturbative approaches. As far as adsorption energies are concerned, pure electrostatic models are in most cases inadequate and simple acid–base models suffer from strong limitations. Finally, it is possible to assess the link between the surface acidity and the structural characteristics of the surface OH groups. The systematic study of the adsorption of more complex molecules on model surfaces should allow a refinement of this analysis in the future.

The analysis performed above shows that the answer to the first question: 'which are the specific properties of the reactants which determine the strength of an acid–base reaction' is not easy. On the one hand, it is clear that (i) the substrate and adsorbate acceptor and donor level positions are relevant for interfacial bond formation, (ii) their effective charges govern the electrostatic interactions and (iii) the cation electronegativity is a key parameter in determining the covalency of the cation–oxygen bond in the substrate and of the interfacial bond, as recognized in the acid–base literature. On the other hand, several examples have been given, which prove that the adsorption characteristics cannot be solely predicted on the basis of the isolated adsorbate and substrate properties. The mutual influence of the adsorbates and substrate is an important factor, which may drive the trends for some series. It shifts the frontier orbital energies – e.g. the case of proton adsorption on MgO(100), (110) and (211) –, and it determines the multiplicity of the bond ($Z_a = 1, 2, 3$) as a function of the reactant geometry and chemical type. It may yield structural distortions in the surface and in the adsorbate. It is especially important when the density of adsorbates is large.

6.6 Conclusion

This chapter was not intended to cover all the results published on surface acid–base properties. The ambition was, rather, to establish some guidelines and put together arguments used in various areas of research. It is true that, in most cases, only qualitative trends are available because most work has been done on materials with direct practical applications – for example, oxide powders or mixed oxides in catalysis – which are far from the model systems necessary to allow a detailed understanding. Additionally, quantum approaches are unlikely in the near future to be able to compete with empirical acid–base approaches in the prediction of adhesion in such applications. Nevertheless, because quantum approaches allow a detailed understanding of model situations, they can give hints on how to monitor the acid strength of surfaces, for example by favouring specific surface orientations, or by introducing selected and controlled densities of structural or chemical defects.

References

Abarenkov I. V., Tret'Yak V. M. and Tulub A. V. (1987) *Sov. J. Chem. Phys.* **4**(7) 1602

Ahn C. C. and Krivanek O. L. (1983) *EELS Atlas* (ASU HREM and GATAN)

Ai M. (1977) *Bull. Chem. Soc. Jpn.* **50** 2579

Akselsen O. M. (1992a) *J. Materials. Sci.* **27** 569

Akselsen O. M. (1992b) *J. Materials. Sci.* **27** 1989

Allan N. L. and Mackrodt W. C. (1994) *Phil. Mag.* **B69** 871

Allan N. L., Cooper D. L. and Mackrodt W. C. (1990) *Molecular Simulations* **4** 269

Anderson A. B., Ravimohan C. and Mehandru S. P. (1987) *Surf. Sci.* **183** 438

Anderson P. W. (1959) *Phys. Rev.* **115** 2

Andersson S. and Davenport J. W. (1978) *Solid State Comm.* **28** 677

Andrews E. H. (1974) *J. Materials Sci.* **9** 887

Argile C. and Rhead G. E. (1989) *Surf. Sci. Rep.* **10** 277

Aryasetiawan F., Gunnarsson O., Knupfer M. and Fink J. (1994) *Phys. Rev.* **B50** 7311

Auroux A. and Gervasini A. (1990) *J. Phys. Chem.* **94** 6371

Baden A. D., Cox P. A., Egdell R. G., Orchard A. F. and Willmer R. J. D. (1981) *J. Phys. C: Solid State Phys.* **14** L1081

Bader R. F. W. (1991) *Chem. Rev.* **91** 893

Baik S., Fowler D. E., Blakely J. M. and Raj R. (1985) *J. Am. Ceram. Soc.* **68** 281

Barbieri A., Weiss W., Van Hove M. A. and Somorjai G. A. (1994) *Surf. Sci.* **302** 259

Bardeen J. (1947) *Phys. Rev.* **71** 717

Barnett R. N. and Bass R. (1979) *Phys. Rev.* **B19** 4259

Barrera R. G. and Duke C. B. (1976) *Phys. Rev.* **B13** 4477

Bart F. and Gautier M. (1994) *Surf. Sci. Lett.* **311** L671

Bart F., Gautier M., Duraud J. P. and Henriot M. (1992) *Surf. Sci.* **274** 317

Bart F., Gautier M., Jollet F. and Duraud J. P. (1994) *Surf. Sci.* **306** 342

Bechstedt F. and Enderlein R. (1979) *Phys. Status Solidi* **B94** 239

Bechstedt F., Del Sole R., Cappellini G. and Reining L. (1992) *Solid State Comm.* **84** 765

Bensoussan M. and Lannoo M. (1979) *J. Physique (France)* **40** 749

Beraud C. and Esnouf C. (1991) *Microsc. Microanal. Microstruct.* **1** 69

Bessières J. and Baro R. (1973) *J. Cryst. Growth* **18** 225

Bickel N., Schmidt G., Heinz K. and Müller K. (1989) *Phys. Rev. Lett.* **62** 2009

Bickel N., Schmidt G., Heinz K. and Müller K. (1990) *Vacuum* **41** 46

Biegelsen D. K., Brigans R. D., Northrup J. E. and Swartz L. E. (1990a) *Phys. Rev. Lett.* **65** 452

Biegelsen D. K., Brigans R. D., Northrup J. E. and Swartz L. E. (1990b) *Phys. Rev.* **B41** 5701

Bilz H. and Kress W. (1979) *Phonon Dispersion Relations in Insulators*, Springer Series in Solid State Science **10** (Springer Verlag: Berlin, Heidelberg, New York)

Binggeli N., Troullier N., Martins J. L. and Chelikowsky J. R. (1991) *Phys. Rev.* **B44** 4771

Binggeli N., Chelikowsky J. R. and Wentzcovitch R. M. (1994) *Phys. Rev.* **B49** 9336

Blanchard D. L., Lessor D. L., La Femina J. P., Baer D. R., Ford W. K. and Guo T. (1991) *J. Vac. Sci. Technol.* **A9** 1814

Blandin A., Castiel D. and Dobrzynski L. (1973) *Solid State Comm.* **13** 1175

Blöchl P., Das G. P., Fishmeister H. F. and Schönberger U. (1990) in *Metal Ceramic Interfaces*, Acta Scr. Met. Proc. Series **4** (Eds.: Rühle M., Evans A. G., Ashby M. F. and Hirth J. P.) (Pergamon Press) p. 9

Blonski S. and Garofalini S. H. (1993) *Surf. Sci.* **295** 263

Boehm H. P. (1971) *Discuss. Faraday Soc.* **52** 264

Boffa A. B., Galloway H. C., Jacobs P. W., Benitez J. J., Batteas J. D., Salmeron M., Bell A. T. and Somorjai G. A. (1995) *Surf. Sci.* **326** 80

Bolger J. C. (1983) in *Adhesion Aspects of Organic Coatings* (Ed.: Mittal K. L.) (Plenum Press: New York) p. 3

Bordier G. and Noguera C. (1991) *Phys. Rev.* **B44** 6361

Bordier G. and Noguera C. (1992) *Le Vide/ Les Couches minces supplément* **260** 268

Born M. and Huang K. (1954) *Dynamic Theory of Crystal Lattices* (Oxford University Press: London)

Bortz M. L. and French R. H. (1989) *Appl. Phys. Lett.* **55** 1955

Boureau G. and Tetot R. (1989) *Phys. Rev.* **B40** 2304

Bourgeois S., Jomard F. and Perdereau M. (1992) *Surf. Sci.* **279** 349

Brand H. V., Curtiss L. A. and Iton L. E. (1992) *J. Phys. Chem.* **96** 7725

Brewis D. M. and Kinloch A. J. (1983) in *Durability of Structural Adhesives* (Applied Science Publishers: London, New York) p. 215

Brillson L. J. (1982) *Surf. Sci. Rep.* **2** 123

Brønsted J. N. (1923) *Rec. Trav. Chim. Pays Bas* **42** 718

Brookes N. B., Law D. S.-L., Padmore T. S., Waeburton D. R. and Thornton G. (1986) *Solid State Comm.* **57** 473

Brookes N. B., Thornton G. and Quinn F. M. (1987) *Solid State Comm.* **64** 383

Broughton J. Q. and Bagus P. S. (1980) *J. Elec. Spect. Relat. Phen.* **20** 261

Brown H. R. (1993) *Macromolecules* **26** 1666

Brown I. D. (1977) *Acta Cryst.* B33 1305

Brusdeylins G., Doak R. B., Skofronick J. G. and Toennies J. P. (1983) *Surf. Sci.* **128** 191

Burggraf L. W., Leyden D. E., Chin R. L. and Hercules D. M. (1982) *J. Catal.* **78** 360

Cappellini G., Del Sole R., Reining L. and Bechstedt F. (1993) *Phys. Rev.* B47 9892

Car R. and Parrinello M. (1985) *Phys. Rev. Lett.* **55** 2471

Carré A. and Schultz J. (1984) *J. Adhesion* **17** 135

Casamassima M., Darque Ceretti E., Etcheberry A. and Aucouturier M. (1991) *Appl. Surf. Sci.* **52** 205

Castanier E. (1995) *Ph.D. thesis*, Orsay (France)

Castanier E. and Noguera C. (1995a) Surf. Sci. accepted

Castanier E. and Noguera C. (1995b) Surf. Sci. accepted

di Castro V. and Polzonetti G. (1987) *Surf. Sci.* **189–190** 1085

di Castro V., Polzonetti G. and Zanoni R. (1985) *Surf. Sci.* **162** 348

di Castro V., Polzonetti G., Ciampi S., Contini G. and Sakho O. (1993) *Surf. Sci.* **293** 41

Catlow C. R. A. and Mackrodt W. C. (1982) *Computer Simulation in Solids*, Lecture Notes in Physics **166** (Springer Verlag: Berlin, Heidelberg, New York)

Catlow C. R. A. and Stoneham A. M. (1983) *J. Phys. C: Solid State Phys.* **16** 4321

Catti M., Valerio G., Dovesi R. and Causà M. (1994) *Phys. Rev.* B49 14179

Catti M., Valerio G. and Dovesi R. (1995) *Phys. Rev.* B51 7441

Causà M. and Pisani C. (1989) *Surf. Sci.* **215** 271

Causà M., Dovesi R., Pisani C. and Roetti C. (1986) *Surf. Sci.* **175** 551

Causà M., Dovesi R., Pisani C. and Roetti C. (1989) *Surf. Sci.* **215** 259

Chabert F. (1992) *Ph.D. thesis*, Grenoble (France)

Chadi D. J. (1979) *Phys. Rev. Lett.* **43** 43

Chang C. C. (1968) *J. Appl. Phys.* **19** 5570

Chang C. C.(1971) *J.Vac. Sci. Technol.* **8** 500

Chang K. J. and Cohen M. L.(1984) *Phys. Rev.* B30 4774

Chatain D., Rivollet I. and Eustathopoulos N. (1986) *J. Chimie Physique* **83** 561

Chatain D., Rivollet I. and Eustathopoulos N. (1987) *J. Chimie. Physique* **84** 201

Chatain D., Ghetta V. and Chabert F. (1993) *Materials Sci. Forum* **126–128** 715

Chelikowsky J. R. and Schlüter M. (1977) *Phys. Rev.* B15 4020

Chelikowsky J. R., Troullier N., Martins J. L. and King H. E. Jr (1991) *Phys. Rev.* B44 489

Chen J. G., Colaianni M. L., Weinberg W. H. and Yates J. T. (1992) *Surf. Sci.* **279** 223

Chen P. J. and Goodman D. W. (1994) *Surf. Sci. Lett.* **312** L767

Chen T. S., de Wette F. W. and Alldredge G. P. (1977) *Phys. Rev.* **B15** 1167

Cheng W. H. and Kung H. H. (1981) *Surf. Sci. Lett.* **102** L21

Chin R. L. and Hercules D. M. (1982) *J. Phys. Chem.* **86** 360

Ching W. Y. (1990) *J. Am. Ceram. Soc.* **73** 3135

Ching W. Y., Gu Z. Q. and Xu Y. N. (1994) *Phys. Rev.* **B50** 1992

Cho K. (1979) *Excitons*, Topics in Current Physics **14** (Springer Verlag: Berlin, Heidelberg, New York)

Chung Y. W., Lo W. J. and Somorjai G. A. (1977) *Surf. Sci.* **64** 588

Cinal M., Del Sole R., Krupski J., Bardyszewski W. and Strinati G. (1987) *Solid State Comm.* **62** 633

Cochran W. (1971) *Crit. Rev. in Solid State* **2** 1

Cohen M. L. (1979) *J. Vac. Sci. Technol.* **16** 1135

Cohen M. L., Lin P. J., Roessler D. M. and Walker W. C. (1967) *Phys. Rev.* **155** 992

Colbourn E. A. (1992) *Surf. Sci. Rep.* **15** 281

Colbourn E. A. and Mackrodt W. C. (1983) *Solid State Ionics* **8** 221

Colbourn E. A., Kendrick J. and Mackrodt W. C. (1983) *Surf. Sci.* **126** 550

Coluccia S., Lavagnino S. and Marchese L. (1988) *Mater. Chem. Phys.* **18** 445

Conard T., Ghijsen J., Vohs J. M., Thiry P. A., Caudano R. and Johnson R. L. (1992) *Surf. Sci.* **265** 31

Conard T., Philippe L., Thiry P. A., Lambin P. and Caudano R. (1993) *Surf. Sci.* **287–288** 382

Condon N. G., Murray P. W., Leibsle F. M., Thornton G., Lennie A. R. and Vaughan D. J. (1994) *Surf. Sci. Lett.* **310** L609

Connell G. and Dumesic J. A. (1986a) *J. Catal.* **101** 103

Connell G. and Dumesic J. A. (1986b) *J. Catal.* **102** 216

Courths R., Noffke J., Wern H. and Heise R. (1990) *Phys. Rev.* **B42** 9127

Coustet V. and Jupille J. (1994a) *Surf. Sci.* **307–309** 1161

Coustet V. and Jupille J. (1994b) *Surf. Interf. Anal.* **22** 280

Cowley A. M. and Sze S. M. (1963) *J. Appl. Phys.* **36** 3212

Cowley J. M. (1986) *Prog. Surf. Sci.* **21** 209

Cox D. F. and Schulz K. H. (1991) *Surf. Sci.* **256** 67

Cox D. F., Fryberger T. B. and Semancik S. (1989) *Surf. Sci.* **224** 121

Cox P. A. (1992) *Transition Metal Oxides: an Introduction to their Electronic Structure and Properties*, International Series of Monographs on Chemistry **27** (Clarendon Press: Oxford)

Cox P. A. and William A. A. (1985) *Surf. Sci.* **152–153** 791

Cox P. A., Egdell R. G. and Naylor P. D. (1983) *J. Elec. Spect. Relat. Phen.* **29** 247

Cui J., Jung D. R. and Frankl D. R. (1990) *Phys. Rev.* **B42** 9701

Cunningham J., Hodnett B. K., Ilyas M., Leahy E. L. and Fierro J. J. G. (1981) *Faraday Discuss. of Chem. Soc.* **72** 283

Czyzyk M. T. and Sawatzky G. A. (1994) *Phys. Rev.* **B49** 14211

Dalmai-Imelik G., Bertolini J. C and Rousseau J. (1977) *Surf. Sci.* **163** 67

Davis D. W. and Rabelais J. W. (1974) *J. Am. Chem. Soc.* **96** 5305

Davis D. W. and Shirley D. A. (1976) *J. Am. Chem. Soc.* **98** 7898

Delamar M. (1990) *J. Elec. Spect. Relat. Phen.* **53** C11

Del Sole R. Reining L. and Godby R. W. (1994) *Phys. Rev.* **B49** 8024

Desjonquères M. C. and Spanjaard D. (1982) *J. Phys. C: Solid State Phys.* **15** 4007

Desjonquères M. C. and Spanjaard D. (1983) *J. Phys. C: Solid State Phys.* **16** 3389

Desjonquères M. C. and Spanjaard D. (1993) *Concepts in Surface Physics*, Surface Science Series **30** (Springer Verlag: Berlin, Heidelberg, New York)

De Vita A., Gillan M. J., Lin J. S., Payne M. C., Stich I. and Clarke L. J. (1992a) *Phys. Rev. Lett.* **68** 3319

De Vita A., Gillan M. J., Lin J. S., Payne M. C., Stich I. and Clarke L. J. (1992b) *Phys. Rev.* **B46** 12964

De Vita A., Manassidis I., Lin J. S. and Gillan M. J. (1992c) *Europhys. Lett.* **19** 605

Dick B. G. and Overhauser A. W. (1958) *Phys. Rev.* **112** 90

Didier F. and Jupille J. (1994a) *Surf. Sci.* **307–309** 587

Didier F. and Jupille J. (1994b) *Surf. Sci.* **314** 378

Diebold U., Pan J. M. and Madey T. E. (1993a) *Surf. Sci.* **287–288** 896

Diebold U., Pan J. M. and Madey T. E. (1993b) *Phys. Rev.* **B47** 3868

Diebold U., Tao H. S., Shinn N. D. and Madey T. E. (1994) *Phys. Rev.* **B50** 14474

Dovesi R., Roetti C. and Freyria-Fava C. (1992) *Phil. Trans. Roy. Soc.* **A341** 203

Drago R. S. and Wayland B. B. (1965) *J. Am. Chem. Soc.* **87** 3571

Duffy D. M., Hoare J. P. and Tasker P. W. (1984) *J. Phys. C: Solid State Physics* **17** L195

Duffy D. M., Harding J. H. and Stoneham A. M. (1992) *Acta Metall. Mater.* **40** S11

Duffy D. M., Harding J. H. and Stoneham A. M. (1993) *Phil. Mag.* **A67** 865

Duffy D. M., Harding J. H. and Stoneham A. M. (1994) *J. Appl. Phys.* **76** 279

Dufour L. C. and Perdereau M. (1988) in *Surface and Near Surface Chemistry of Oxide Materials*, Materials Science Monographs **47** (Eds.: Nowotny J. and Dufour L. C.) (Elsevier: Amsterdam), p. 577

Duke C. B., Meyer R. J., Paton A. and Mark P. (1978) *Phys. Rev.* **B18** 4225

Ealet B. (1993) *Ph.D. thesis*, Marseille (France)

Ealet B., Gillet E., Nehasil V. and Møller P. J. (1994) *Surf. Sci.* **318** 151

Eastman J. A. and Rühle M. (1989) *Ceramic Engineering and Sci. Proceedings* **10** 1515

Egdell R. G., Eriksen S. and Flavell W. R. (1986) *Solid State Comm.* **60** 835

Ellialtioglu S., Wolfram T. and Henrich V. E. (1978) *Solid State Comm.* **27** 321

Elliott R. J. (1957) *Phys. Rev.* **108** 1384

Epicier T., Esnouf C., Smith M. A. and Pope D. (1993) *Phil. Mag. Lett.* **65** 299

Ernst H. J., Hulpke and Toennis J. P. (1987) *Phys. Rev. Lett.* **58** 1941

Ernst K. H., Ludviksson A., Zhang R., Yoshihara J. and Campbell C. T. (1993) *Phys. Rev.* **B47** 13782

Eustathopoulos N. and Drevet B. (1994) *J. Physique (France) III* **4** 1865

Eustathopoulos N., Chatain D. and Couturier L. (1991) *Materials Sci. Ing.* **A135** 83

Evarestov R. A. and Veryazov V. A. (1990) *Phys. Status Solidi* **B158** 201

Evarestov R. A. and Veryazov V. A. (1991) *Phys. Status Solidi* **B165** 411

Evarestov R. A., Shapovalov V. O. and Veryazov V. A. (1994) *Phys. Status Solidi* **B183** K15

Evjen H. M. (1932) *Phys. Rev.* **39** 675

Ewald P. P. (1921) *Ann. Physik* **64** 253

Fahmi A. and Minot C. (1994) *Surf. Sci.* **304** 343

Fasolino A., Santoro G. and Tosatti E. (1980) *Phys. Rev. Lett.* **44** 1684

Felton R. C., Prutton M., Tear S. P. and Welton-Cook M. R. (1979) *Surf. Sci.* **88** 474

Feuchtwang T. E., Paudyal D. and Pong W. (1982) *Phys. Rev.* **B26** 1608

Finnis M. W., Stoneham A. M. and Tasker P. W. (1990) in *Metal Ceramic Interfaces*, Acta Scr. Met. Proc. Series **4** (Eds.: Rühle M., Evans A. G., Ashby M. F. and Hirth J. P.) (Pergamon Press) p. 35

Finston H. L. and Rychtman A. C. (1982) *A New View of Current Acid–Base Theories* (J. Wiley and Sons, Inc.: New York)

Floquet N. and Dufour L. C. (1983) *Surf. Sci.* **126** 543

Flores F. and Tejedor C. (1987) *J. Phys. C: Solid State Phys.* **20** 145

Fowkes F. M. (1987) *J. Adhesion Sci. Technol.* **1** 7

Fowkes F. M. (1990) *J. Adhesion Sci. Technol.* **4** 669

Fowler P. W. and Tole P. (1988) *Surf. Sci.* **197** 457

Frank F. C. (1950) *Phil. Mag.* **41** 1287

Frederick B. G., Apai G. and Rhodin T. N. (1991a) *Phys. Rev.* **B44** 1880

Frederick B. G., Apai G. and Rhodin T. N. (1991b) *Surf. Sci.* **244** 67

Frederick B. G., Apai G. and Rhodin T. N. (1992) *Surf. Sci.* **277** 337

Freeman A. J., Li C. and Fu C. L. (1990) in *Metal Ceramic Interfaces*, Acta Scr. Met. Proc. Series **4** (Eds.: Rühle M., Evans A. G., Ashby M. F. and Hirth J. P.) (Pergamon Press) p. 2

French T. M. and Somorjai G. A. (1970) *Phys. Chem.* **74** 2489

French R. H., Kasowski R. V., Ohuchi F. S., Jones D. J., Song H. and Coble R. L. (1990) *J. Am. Ceram. Soc.* **73** 3195

Freyria-Fava C., Dovesi R., Saunders V. R., Leslie M. and Roetti C. (1993) *J Phys. Condensed Matter* **5** 4793

Friedel J. and Noguera C. (1983) *Int. J. Quantum Chem.* **23** 1209

Fuchs R. and Kliewer K. L. (1965) *Phys. Rev.* **140** A2076

Gajdardziska-Josifovska M., Crozier P. A. and Cowley J. M. (1991) *Surf. Sci Lett.* **248** L259

Galloway H. C., Benitez J. J. and Salmeron M. (1993) *Surf. Sci.* **298** 127

Gan J. L., Ong H. W. K., Tan T. L. and Chan S. O. (1986) *J. Appl. Polymer Sci.* **32** 4109

Gao Y., Shewmon P. and Dregia A. (1988) *Scripta Metall.* **22** 1521

Garcia A. and Cohen M. L. (1993) *Phys. Rev.* B47 4215

Garrone E., Zecchina A. and Stone F. S. (1980) *Phil. Mag.* B42 683

Gaspard J. P. and Cyrot-Lackmann F. (1973) *J. Phys. C: Solid State Phys.* **6** 3077

Gassman P., Franchy R. and Ibach H. (1994) *Surf. Sci.* **319** 95

Gautier M. and Duraud J. P. (1994) *J. Physique (France) III* **4** 1779

Gautier M., Duraud J. P. and Pham Van L. (1991a) *Surf. Sci. Lett.* **249** L327

Gautier M., Duraud J. P., Pham Van L. and Guittet M. J. (1991b) *Surf. Sci.* **250** 71

Gautier M., Renaud G., Pham Van L., Villette B., Pollak M., Thromat N., Jollet F. and Duraud J. P. (1994) *J. Am. Ceram. Soc.* **77** 323

de Gennes P. G. (1989) *J. Physique (France)* **50** 2551

Gent A. N. and Schultz J. (1972) *J. Adhesion* **3** 281

Gercher V. A. and Cox D. F. (1995) *Surf. Sci.* **322** 177

Gervasini A. and Auroux A. (1991) *J. Catal.* **131** 190

Gervasini A., Bellussi G., Fenyvesi J. and Auroux A. (1993) in *New Frontiers in Catalysis* (Ed.: Guczi L.) (Elsevier: Amsterdam)

Gibson A., Haydock R. and LaFemina J. P. (1992) *J. Vac. Sci. Technol.* **A10** 2361

Gillan M. J., Manassidis I. and De Vita A. (1994) *Phil. Mag.* B69 879

Gillet E. and Ealet B. (1992) *Surf. Sci.* **273** 427

Gillet E., Legressus C. and Gillet M. F. (1987) *J. Chim. Phys.* **84** 167

Gillis P. P. and Gilman J. J. (1964) *J. Appl. Phys.* **35** 647

Giorgio S., Henry C. R., Chapon C. and Penisson J. M. (1990) *J. Cryst. Growth* **100** 254

Giorgio S., Chapon C., Henry C. R. and Nihoul G. (1993) *Phil. Mag.* B67 773

Glassford K. M., Troullier N., Martins J. L. and Chelikowsky J. R. (1990) *Solid State Comm.* **76** 635

Godby R. W., Schlüter M. and Sham L. J. (1988) *Phys. Rev.* B37 10159

Godin T. J. and LaFemina J. P. (1993) *Phys. Rev.* B47 6518

Godin T. J. and LaFemina J. P. (1994a) *Phys. Rev.* B49 7691

Godin T. J. and LaFemina J. P. (1994b) *Surf. Sci.* **301** 364

Goniakowski J. (1994) *Ph.D. thesis*, Orsay (France)

Goniakowski J. and Noguera C. (1993) *Le Vide/Les Couches minces, supplément* **268** 11

Goniakowski J. and Noguera C. (1994a) *Surf. Sci.* **319** 68

Goniakowski J. and Noguera C. (1994b) *Surf. Sci.* **319** 81

Goniakowski J. and Noguera C. (1994c) *Le Vide/Les Couches minces, supplément* **272** 74

Goniakowski J. and Noguera C. (1995a) *Surf. Sci.* **323** 129

Goniakowski J. and Noguera C. (1995b) *Surf. Sci.* **330** 337

Goniakowski J. and Noguera C. (1995c) *Surf. Sci.* **340** 191

Goniakowski J., Russo S. and Noguera C. (1993) *Surf. Sci.* **284** 315

Göpel W., Pollmann J., Ivanov I. and Reihl B. (1982) *Phys. Rev.* B**26** 3144

Göpel W., Anderson J. A., Frankel D., Jaehnig M., Phillips K., Schäfer J. A. and Rocker G. (1984) *Surf. Sci.* **139** 333

Gota S., Gautier M., Douillard L., Thromat N., Duraud J. P. and Le Fèvre P. (1995) *Surf. Sci.* **323** 163

Goyhenex C. and Henry C. R. (1992) *J. Elec. Spect. Relat. Phen.* **61** 65

Grabow M. H. and Gilmer G. H. (1988) *Surf Sci.* **194** 333

Gray D. E. (1963) *American Institute of Physics Handbook*, 2nd edition

Gu Z. Q. and Ching W. Y. (1994) *Phys. Rev.* B**49** 13958

Guo J., Ellis D. E. and Lam D. J. (1992) *Phys. Rev.* B**45** 13647

Gupta R. P. (1985) *Phys. Rev.* B**32** 8278

Gutmann V. (1978) *The Donor–Acceptor Approach to Molecular Interactions* (Plenum Press: New York, London)

Gutzwiller M. C. (1965) *Phys. Rev.* **137** A1726

Hair M. L. and Hertl W. (1970) *J. Phys. Chem.* **74** 91

Hannemann H., Ventrice Jr. C. A., Bertrams T., Brodde A. and Neddermeyer H. (1994) *Phys. Status Solidi* A**146** 289

Harding J. H. (1990) *Rep. Prog. Phys.* **53** 1403

Harding J. H. (1991) in *Computer Simulation in Material Science*, NATO ASI series **205** (Eds.: Mayer M. and Pontikis V.), p. 159

Harding J. H. (1992) *Phil. Trans. Roy. Soc.* **341** 283

Hardman P. J., Prakash N. S., Muryn C. A., Raikar G. N., Thomas A. G., Prime A. F. and Thornton G. (1993) *Phys. Rev.* B**47** 16056

Hardman P. J., Raikar G. N., Muryn C. A., van der Laan G., Wincott P. L., Thornton G., Bullett D. W. and Dale P. A. D. M. A. (1994) *Phys. Rev.* B**49** 7170

Harrison N. M. and Leslie M. (1992) *Molecular Simulation* **9** 171

Harrison W. A. (1980) *Electronic Structure and the Properties of Solids* (W. H. Freeman and company: San Francisco, CA)

Harrison W. A. (1981) *Phys. Rev.* B**23** 5230

Harrison W. A. (1985) *Phys. Rev.* B**31** 2121

Hartman P. (1980) *J. Cryst. Growth* **49** 166

Hartman P. and Bennema P. (1980) *J. Cryst. Growth* **49** 145

Hartman P and Perdock W. (1955) *Acta Cryst.* **8** 49, 521, 525

Hassel M. and Freund H. J. (1995) *Surf. Sci.* **325** 163

Haydock R., Heine V. and Kelly M. J. (1972) *J. Phys. C: Solid State Phys.* **5** 2845

Haydock R., Heine V. and Kelly M. J. (1975) *J. Phys. C: Solid State Phys.* **8** 2591

He J. W. and Møller P. J. (1986) *Phys. Stat. Sol.* B**133** 687

Hemstreet L. A., Chubb S. R. and Pickett W. E. (1988) *J. Vac. Sci. Technol.* A**6** 1067

Henderson M. A. (1994) *Surf. Sci.* **319** 315

Hengehold R. L. and Pedrotti F. L. (1976) *J. Appl. Phys.* **47** 287

Henrich V. E. (1976) *Surf. Sci.* **57** 385

Henrich V. E. (1983) *Prog. Surf. Sci.* **14** 175

Henrich V. E. (1985) *Rep. Prog. Phys.* **48** 1481

Henrich V. E. and Cox P. A. (1994) *The Surface Science of Metal Oxides* (Cambridge University Press: Cambridge)

Henrich V. E. and Kurtz R. L. (1981a) *Phys. Rev.* **B23** 6280

Henrich V. E. and Kurtz R. L. (1981b) *J. Vac. Sci. Technol.* **18** 416

Henrich V. E., Dresselhaus G. and Zeiger H. J. (1976) *Phys. Rev. Lett.* **36** 1335

Henrich V. E., Dresselhaus G. and Zeiger H. J. (1978) *Phys. Rev.* **B17** 4908

Henrich V. E., Dresselhaus G. and Zeiger H. J. (1980) *Phys. Rev.* **B22** 4764

Henry C. R. and Poppa H. (1990) *Thin Solid Films* **189** 303

Henry C. R., Chapon C., Duriez C. and Giorgio S. (1991) *Surf. Sci.* **253** 177

Henzler M. (1973) *Surf. Sci.* **36** 109

Hiemstra T., van Riemsdijk W. H. and Bolt G. H. (1989a) *J. Colloid and Interf. Sci.* **133** 91

Hiemstra T., de Wit J. C. M. and van Riemsdijk W. H. (1989b) *J. Colloid and Interf. Sci.* **133** 105

Hikita T., Hanada T. and Kudo M. (1993) *Surf. Sci.* **287–288** 377

Hirata A., Ando A., Saiki K. and Koma A. (1994a) *Surf. Sci.* **310** 89

Hirata A., Saiki K., Koma A. and Ando A. (1994b) *Surf. Sci.* **319** 267

Hoffmann R. (1963) *J. Chem. Phys.* **39** 1397

Hong S. Y., Anderson A. B. and Smialek J. L. (1990) *Surf. Sci.* **230** 175

Höss C., Wolfe J. P. and Kinder H. (1990) *Phys. Rev. Lett.* **64** 1134

Hott R. (1991) *Phys. Rev.* **B44** 1057

Hsu T. and Kim Y. (1991) *Surf. Sci.* **258** 119

Hubbard J. (1964) *Proc. Roy. Soc.* **A281** 401

Hudson L. T., Kurtz R. L., Robey S. W., Temple D. and Stockbauer R. L. (1993a) *Phys. Rev.* **B47** 1174

Hudson L. T., Kurtz R. L., Robey S. W., Temple D. and Stockbauer R. L. (1993b) *Phys. Rev.* **B47** 10832

Hudson R. F. and Klopman G. (1967) *Tetrahedron Lett.* **12** 1103

Hüfner S., Steiner P., Sander I., Reinert F. and Schmitt H. (1992) *Z. Phys.* **B86** 207

Hugenschmidt M. B., Gamble L. and Campbell C. T. (1994) *Surf. Sci.* **302** 329

Hybertsen M. S. and Louie S. G. (1986) *Phys. Rev.* **B34** 5390

Hybertsen M. S. and Louie S. G. (1987a) *Phys. Rev.* **B35** 5585

Hybertsen M. S. and Louie S. G. (1987b) *Phys. Rev.* **B35** 5602

Ibach H. J. (1970) *Phys. Rev. Lett.* **24** 1416

Ibach H. J. (1972) *J. Vac. Sci. Technol.* **9** 713

Iczkowski R. P. and Margrave J. L. (1961) *J. Am. Chem. Soc.* **83** 3547

Inglesfield J. E. and Wikborg E. (1975) *J. Phys. F: Metal Phys.* **5** 1475

Ishida H. and Koenig J. L. (1980) *J. Polymer Sci.* **18** 233

Ivanov I. and Pollmann J. (1980) *Solid State Comm.* **36** 361

Ivanov I. and Pollmann J. (1981a) *J. Vac. Sci. Technol.* **19** 344

Ivanov I. and Pollmann J. (1981b) *Phys. Rev.* **B24** 7275

Jensen W. B. (1978) *Chem. Rev.* **78** 1

Jensen W. B. (1980) *The Lewis Acid–Base Concepts: an Overview* (Wiley-Interscience: New York)

Jensen W. B. (1991) in *Acid–Base Interactions: Relevance to Adhesion Science and Technology* (Eds.: Mittal K. L. and Anderson H. R., Jr) (VSP BV: Utrecht, The Netherlands) p. 3

Johnson K. H. and Pepper S. V. (1982) *J. Appl. Phys.* **53** 6634

Jollet F., Thromat N., Gautier M., Duraud J. P. and Noguera C. (1992) *Physica Scripta* **41** 251

Jortner J. (1992) *Z. Phys.* **D24** 247

Joyes P. (1990) *Les Agrégats Inorganiques Elémentaires*, Monographies de Physique (Les Editions de Physique: Les Ulis, France)

Julien J. P., Mayou D. and Cyrot-Lackmann F. (1989) *J. Physique (France)* **50** 2683

Jung D. R., Cui J. and Frankl D. R. (1991) *J. Vac. Sci. Technol.* **A9** 1589

Jura G. and Garland C. W. (1952) *J. Am. Chem. Soc.* **74** 6033

Kantorovich L. N., Holender J. M. and Gillan M. J. (1995) *Surf. Sci.* accepted

Kasowski R. V. and Tait R. H. (1979) *Phys. Rev.* **B20** 5168

Kasowski R. V., Ohuchi F. S. and French R. H. (1988) *Physica* **B150** 44

Kassim H. A., Matthew J. A. D. and Green B. (1978) *Surf. Sci.* **74** 109

Kesmodel L. L., Gates J. A. and Chung Y. W. (1981) *Phys. Rev.* **B23** 489

Kendall K. (1971) *J. Phys. D: Appl. Phys.* **4** 1186

Kendall K. (1975) *J. Phys. D: Appl. Phys.* **8** 1449

Kim S. S., Baik S., Kim H. W. and Kim C. Y. (1993) *Surf. Sci. Lett.* **294** L935

Kim Y. and Hsu T. (1991) *Surf. Sci.* **258** 131

Kinniburgh C. G. (1975) *J. Phys. C: Solid State Phys.* **8** 2832

Kinniburgh C. G. and Walker J. A. (1977) *Surf. Sci.* **63** 274

Kittel C. (1990) *Physique de l'État Solide*, 5ème édition (Dunod Université)

Klopman G. and Hudson R. F. (1967) *Theor. Chim. Acta* **8** 165

Knözinger E., Jacob K. H. and Hofmann P. (1993a) *J. Chem. Soc. Faraday Trans.* **89** 1101

Knözinger E., Jacob K. H., Singh S. and Hofmann P. (1993b) *Surf. Sci.* **290** 388

Kowalczyk S. P., McFeely F. R., Ley L., Gritsyna V. T. and Shirley D. A. (1977) *Solid State Comm.* **23** 161

Kress W. and de Wette F. W. (1991) *Surface Phonons*, Springer Series in Surface Science **21** (Springer Verlag: Berlin, Heidelberg, New York)

Kurtin S., McGill T. C. and Mead C. A. (1969) *Phys. Rev. Lett.* **22** 1433

Kurtz R. L., Stockbauer R., Madey T. E., Roman E. and De Segovia J. L. (1989) *Surf. Sci.* **218** 178

Lad R. J. and Henrich V. E. (1988) *Surf. Sci.* **193** 81

LaFemina J. P (1992) *Surf. Sci. Rep.* **16** 133

LaFemina J. P. and Duke C. B. (1991) *J. Vac. Sci. Technol.* A**9** 1847

Lake G. and Thomas A. (1967) *Proc. Roy. Soc.* A**300** 108

Lakshmi G. and de Wette F. W. (1980) *Phys. Rev.* B**22** 5009

Lambin P., Vigneron J. P. and Lucas A. A. (1985) *Phys. Rev.* B**32** 8203

Landau L. and Lifchitz E. (1967) *Théorie de l'Elasticité* (Editions Mir: Moscou)

Lang N. D. and Kohn W. (1970) *Phys. Rev.* B**1** 4555

Langel W. and Parrinello M. (1994) *Phys. Rev. Lett.* **73** 504

Lannoo M. (1993) *Materials Sci. Ing.* B**22** 1

Lannoo M. and Friedel P. (1991) *Atomic and Electronic Structure of Surfaces*, Springer Series In Surface Science **16** (Springer Verlag: Berlin, Heidelberg, New York)

Laurent V. (1988) *Ph.D. thesis*, Grenoble (France)

Lawrence P. J., Parker S. C. and Tasker P. W. (1988) *J. Am. Ceram. Soc.* C**71** 389

Levine J. D. and Mark P. (1966) *Phys. Rev.* **144** 751

Levine Z. H. and Louie S. G. (1982) *Phys. Rev.* B**25** 6310

Lewis G. N. (1923) *Valence and the Structure of Atoms and Molecules* (The Chemical Catalog Co.: New York)

Li C., Wu R., Freeman A. J. and Fu C. L. (1993) *Phys. Rev.* B**48** 8317

Li X., Liu L. and Henrich V. E. (1992) *Solid State Comm.* **84** 1103

Liang Y. and Bonnell D. A. (1993) *Surf. Sci. Lett.* **285** L510

Liang Y. and Bonnell D. A. (1994) *Surf. Sci.* **310** 128

Liehr M., Thiry P. A., Pireaux J. J. and Caudano R. (1984) *J. Vac. Sci. Technol.* A**2** 1079

Liehr M., Thiry P. A., Pireaux J. J. and Caudano R. (1986) *Phys. Rev.* B**33** 5682

Lin X. Y and Arribart H. (1993) *Le Vide/Les Couches minces, supplément* **268** 31

Lin X. Y., Creuzet F. and Arribart H. (1993) *J. Phys. Chem.* **97** 7272

Lindan P. J. D. and Gillan M. J. (1994) *Phil. Mag.* B**69** 535

von der Linden W. and Horsh P. (1988) *Phys. Rev.* B**37** 8351

Lindqvist I. (1963) *Inorganic Adduct Molecules of Oxo-Compounds* (Springer Verlag: Berlin, Göttingen, Heidelberg)

Liu F., Garofalini S. H., King-Smith D. and Vanderbilt D. (1994) *Phys. Rev.* B**49** 12528

Livey D. T. and Murray P. (1956) *J. Am. Ceram. Soc.* **39** 363

Lo W. J. and Somorjai G. A. (1978) *Phys. Rev.* B**17** 4942

Lo W. J., Chung Y. W. and Somorjai G. A. (1978) *Surf. Sci.* **71** 199

Louie S. G. and Cohen M. L. (1975) *Phys. Rev. Lett.* **35** 866

Louie S. G., Chelikowskiy J. R. and Cohen M. L. (1976) *J. Vac. Sci. Technol.* **13** 790

Louie S. G., Chelikowsky J. R. and Cohen M. L. (1977) *Phys. Rev.* B**15** 2154

Lowry T. M. (1923a) *Chem. Ind. (London)* **42** 43

Lowry T. M. (1923b) *Chem. Ind. (London)* **42** 1048

Lubinsky A. R., Duke C. B., Chang S. C., Lee B. W. and Mark P. (1976) *J. Vac. Sci. Technol.* **13** 189

Lucas A. A. (1968) *J. Chem. Phys.* **48** 3156

Lucas A. A. and Vigneron J. P. (1984) *Solid State Comm.* **49** 327

Mackrodt W. C (1984) *Solid State Ionics* **12** 175

Mackrodt W.C (1988) *Phys. Chem. Miner.* **15** 228

Mackrodt W. C (1989) *J. Chem. Soc. Faraday Trans II.* **85** 54

Mackrodt W. C and Stewart R. F. (1977) *J. Phys. C: Solid State Phys.* **10** 1431

Mackrodt W. C and Tasker P. W. (1985) *Chem. Britain* **21** 13

Mackrodt W. C., Davey R. J., Black S. N. and Docherty R. (1987) *J. Cryst. Growth* **80** 441

Mackrodt W. C., Harrison N. M., Saunders V. R., Allan N. L., Towler M. D., Apra E. and Dovesi R. (1993) *Phil. Mag.* A**68** 653

Mader W. and Maier B. (1990) *J. Physique (France)* **51** C1 867

Mader W. and Rühle M. (1989) *Acta Metall. Mater.* **37** 853.

Majewski J. A. and Vogl P. (1986) *Phys. Rev. Lett.* **57** 1366

Maksym P. A. (1985) *Surf. Sci.* **149** 157

Manassidis I., De Vita A. and Gillan M. J. (1993) *Surf. Sci. Lett.* **285** L517

Manassidis I., Goniakowski J., Kantorovich L. N. and Gillan M. J. (1995) *Surf. Sci.* accepted

Martin A. J. and Bilz H. (1979) *Phys. Rev.* B**19** 6593

Martin R. L. and Shirley D. A. (1974) *J. Am. Chem. Soc.* **96** 5299

Martins T. P. (1983) *Phys. Rep.* **95** 167

Maschhoff B. L., Pan J. M. and Madey T. E. (1991) *Surf. Sci.* **259** 190

Massida V. (1978) *Physica* **95**B 317

Matsumoto T., Tanaka H., Kawai T. and Kawai S. (1992) *Surf. Sci. Lett.* **278** L153

Mattheiss L. F. (1972) *Phys. Rev.* B**6** 4718

Maugis D. (1977) *Le Vide/Les Couches minces* **186** 1

Maurice V., Salmeron M. and Somorjai G. A. (1990) *Surf. Sci.* **237** 116

McDonald J. E. and Eberhart J. G. (1965) *Trans. Met. Soc. AIME* **233** 512

van Meerssche M. and Feneau-Dupont J. (1977) *Introduction à la Cristallographie et à la Chimie Structurale*, supplément à la 2ème édition (Oyez Editeur: Louvain, Bruxelles, Paris)

Mele E. J. and Joannopoulos J. D. (1978) *Phys. Rev.* B**17** 1528

Mittal K. L. and Anderson H. R., Jr. (1991) *Acid–Base Interactions: Relevance to Adhesion Science and Technology* (VSP BV: Utrecht, The Netherlands)

Miyazaki T., Kuroda Y., Morishige K., Kittaka S., Umemura J., Takenaka T. and Morimoto T. (1985) *J. Colloid and Interf. Sci.* **106** 154

Møller P. J. and Guo Q. (1991) *Thin Solid Films* **201** 267

Møller P. J. and Nerlov J. (1994) *Surf. Sci.* **307–309** 591

Moon A. R. and Phillips M. R. (1991) *J. Am. Ceram. Soc.* **74** 865

Morrow B. A. (1990) *Stud. Surf. Sci. Catal.* A**57** 161

Mott N. F. (1974) *Metal–Insulator Transitions* (Taylor and Francis, Ltd.: London)

Mott N. F. and Gurney R. W. (1948) *Electronic Processes in Ionic Crystals*, 2nd edition (Clarendon Press: Oxford)

Moukouri S. and Noguera C. (1992) *Z. Phys.* D24 71

Moukouri S. and Noguera C. (1993) *Z. Phys.* D27 79

Mudler C. A. M. and Klomp J. T. (1985) *J. Physique (France)* 46 C4 111

Mueller D. R., Kurtz R. L., Stockbauer R. L., Madey T. E. and Shih A. (1990) *Surf. Sci.* 237 72

Mulheran P. A. and Harding J. H. (1993) *J. Phys. (France) Colloques* 3 1971

Mulliken R. S. (1951) *J. Chem. Phys.* 19 514

Mulliken R. S. (1952a) *J. Chem. Phys.* 56 801

Mulliken R. S. (1952b) *J. Am. Chem. Soc.* 74 811

Mullins W. M. and Averbach B. L. (1988) *Surf. Sci.* 206 41

Munnix S. and Schmeits M. (1985) *Solid State Comm.* 43 867

Murray P. W., Leibsle F. M., Muryn C. A., Fischer H. J., Flipse C. F. J. and Thornton G. (1994a) *Surf. Sci.* 321 217

Murray P. W., Leibsle F. M., Muryn C. A., Fischer H. J., Flipse C. F. J. and Thornton G. (1994b) *Phys. Rev. Lett.* 72 689

Murray P. W., Condon N. J. and Thornton G. (1995) *Surf. Sci. Lett.* 323 L281

Muryn C. A., Hardman P. J., Crouch J. J., Raiker G. N., Thornton G. and Law D. S.-L. (1991) *Surf. Sci.* 251–252 747

Myers R. T. (1974) *Inorg. Chem.* 13 2040

Nabavi M., Spalla O. and Cabanne B. (1993) *J. Colloid and Interf. Sci.* 160 459

Naidich J. V. (1981) *Prog. Surf. Memb. Sci.* 14 353

Nakamatsu H., Sudo A. and Kawai S. (1988) *Surf. Sci.* 194 265

Nakamatsu H., Sudo A. and Kawai S. (1989) *Surf. Sci.* 223 193

Nassir M. H. and Langell M. A. (1994) *Solid State Comm.* 92 791

Nath K. and Anderson A. B. (1989) *Phys. Rev.* B39 1013

Noguera C. and Bordier G. (1994) *J. Physique (France)* III 4 1851

Noguera C., Goniakowski J. and Bouette-Russo S. (1993) *Surf. Sci.* 287–288 188

Noller H., Lercher J. A. and Vinek H. (1988) *Materials Chem. Phys.* 18 577

Nollery H. and Ritter G. (1984) *J. Chem. Soc. Faraday Trans. I* 80 275

Nosker R. W., Mark P. and Levine J. D. (1970) *Surf. Sci.* 19 291

Ohuchi F. S. and Koyama M. (1991) *J. Am. Ceram. Soc.* 74 1163

Oliver P. M., Parker S. C., Purton J. and Bullett D. W. (1994) *Surf. Sci.* 307–309 1200

Onishi H. and Iwasawa Y. (1994) *Surf. Sci. Lett.* 313 L783

Onishi H., Egawa C., Aruga T. and Iwasawa Y. (1987) *Surf. Sci.* 191 479

Onishi H., Aruga T. and Iwasawa Y. (1994) *Surf. Sci.* 310 135

Orlando R., Pisani C., Ruiz E. and Sautet P. (1992) *Surf. Sci.* 275 482

Orlando R., Dovesi R., Roetti C. and Saunders V. R. (1994) *Chem. Phys. Lett* 275 482

Overbury S. H., Bertrand P. A. and Somorjai G. A. (1975) *Chem. Rev.* 75 547

Pacchioni G. and Bagus P. S. (1994) *Phys. Rev.* B50 2576

Pacchioni G., Minerva T. and Bagus P. S. (1992a) *Surf. Sci.* **275** 450

Pacchioni G., Bagus P. S. and Parmigiani F. (1992b) in *Cluster Models for Surface and Bulk Phenomena*, NATO ASI series **283** (Plenum: New York)

Pacchioni G., Sousa C., Illas F., Parmigiani F. and Bagus P. S. (1993) *Phys. Rev.* **B48** 11573

Pak V. N. (1974) *Russian J. Phys. Chem.* **48** 969

Palmberg P. W., Rhodin T. N. and Todd C. J. (1967) *Appl. Phys. Lett.* **11** 33

Pan J. M. and Madey T. E. (1993) *J. Vac. Sci. Technol.* **A11** 1667

Pan J. M., Maschhoff B. L., Diebold U. and Madey T. E. (1992) *J. Vac. Sci. Technol.* **A10** 2470

Pan J. M., Diebold U., Zhang L. and Madey T. E. (1993) *J. Surf. Sci.* **295** 411

Pantelides S. T. and Harrison W. A. (1976) *Phys. Rev.* **B13** 2667

Parks G. A. (1965) *Chem. Rev.* **65** 177

Parr R. G. and Pearson R. G. (1983) *J. Am. Chem. Soc.* **105** 7512

Parr R. G. and Yang W. (1986) *J. Am. Chem. Soc.* **106** 4049

Parry D. E. (1975) *Surf. Sci.* **49** 433

Pauling L. (1929) *J. Am. Chem. Soc.* **51** 1010

Pauling L. (1960) *The Nature of the Chemical Bond*, 3rd edition (Cornell: Ithaca, NY) chapter 13

Pauling L. (1971) *Physics Today* p. 9

Pearson R. G. (1963) *J. Am. Chem. Soc.* **85** 3533

Pearson R. G. (1966) *Science* **151** 172

Pearson R. G. (1989) *J. Org. Chem.* **54** 1423

Pelmenschikov A. G., Morosi G. and Gamba A. (1991) *J. Phys. Chem.* **95** 10037

Penn D. R. (1962) *Phys. Rev.* **128** 2093

Pézerat H. (1989) in *Les Pneumopathies Professionnelles* (Ed.: P. Sebastien) (Editions de l'INSERM: Paris)

Pham Van L., Gautier M., Duraud J. P., Gillet F. and Jollet F. (1990) *Surf. Interf. Anal.* **16** 214

Phillips J. C. (1970) *Rev. Mod. Phys.* **42** 317

Pick S. (1990) *Surf. Sci. Rep.* **12** 99

Pisani C., Dovesi R. and Roetti C. (1992) *Hartree–Fock Ab Initio Treatment of Crystalline Systems*, Lecture Notes in Chemistry **48** (Springer Verlag: Berlin, Heidelberg, New York)

Piveteau B. and Noguera C. (1991) *Phys. Rev.* **B43** 493

Pollak M. (1995) *Ph.D. thesis*, Paris (France)

Pong W. and Paudyal D. (1981) *Phys. Rev.* **B23** 3085

Pople J. A. and Segal G. A. (1965a) *J. Chem. Phys.* **43** 5136

Pople J. A. and Segal G. A. (1965b) *J. Chem. Phys.* **44** 3289

Pople J. A., Sentry D. P. and Segal G. A. (1965a) *J. Chem. Phys.* **43** 5129

Pople J. A., Sentry D. P. and Segal G. A. (1965b) *J. Chem. Phys.* **47** 158

Poppa H., Papageorgopoulos C. A., Marks F. and Bauer E. (1986) *Z. Phys.* **D3** 279

Powel R. J. and Spicer W. E. (1970) *Phys. Rev.* **B2** 2185

Prade J., Schröder U., Kress W., de Wette F. W. and Kulkarni A. D (1993) *J. Phys: Condensed Matter* **5** 1

Priester C., Allan G. and Lannoo M. (1988) *J. Vac. Sci. Technol.* **B6** 1290

Protheroe A. R., Steinbrunn A. and Gallon T. E. (1982) *J. Phys. C: Solid State Phys.* **15** 4951

Protheroe A. R., Steinbrunn A. and Gallon T. E. (1983) *Surf. Sci.* **126** 534

Prutton M., Ramsey J. A., Walker J. A. and Welton-Cook M. R. (1979) *J. Phys. C: Solid State Phys.* **12** 5271

Purton J., Jones R., Catlow C. R. A. and Leslie M. (1993) *Phys. Chem. Miner.* **19** 392

Quiu S. L., Pan X., Strongin M. and Citrin P. H. (1987) *Phys. Rev.* **B36** 1292

Raikar G. N., Hardman P. J., Muryn C. A., van der Laan G., Wincott P. L., Thornton G. and Bullett D. W. (1991) *Solid State Comm.* **80** 423

Ramamoorthy M., King-Smith R. D. and Vanderbilt D. (1994) *Phys. Rev.* **B49** 7709

Rayleigh L. (1885) *Proc. London Math. Soc.* **17** 4

Reining L. and Del Sole R. (1991a) *Phys. Rev.* **B44** 12918

Reining L. and Del Sole R. (1991b) *Phys. Rev. Lett.* **67** 3816

Renaud R., Villette B., Vilfan I. and Bourret A. (1994) *Phys. Rev. Lett.* **73** 1825

Rhoderick E. H. (1978) *Metal–Semiconductor Contacts* (Clarendon: Oxford)

Rohr F., Wirth K., Libuda J., Cappus D., Baümer M. and Freund H. J. (1994) *Surf. Sci. Lett.* **315** L977

Röhrer G. S. and Bonnell D. A. (1991) *Surf. Sci. Lett.* **247** L195

Röhrer G. S., Henrich V. E. and Bonnell D. A. (1990) *Science* **250** 1239

Röhrer G. S., Henrich V. E. and Bonnell D. A. (1992) *Surf. Sci.* **278** 146

Rose H., Smith J. R. and Ferrante J. (1983) *Phys. Rev.* **B28** 1835

Russo S. and Noguera C. (1992a) *Surf. Sci.* **262** 245

Russo S. and Noguera C. (1992b) *Surf. Sci.* **262** 259

Sakata K., Honma K., Ogawa K., Watanabe O. and Nii K. (1986) *J. Materials Sci.* **21** 4463

Sambi M., Granozzi G., Rizzi G. A., Casarin M. and Tondello E. (1994) *Surf. Sci.* **319** 149

Sander M. and Engel T. (1994) *Surf. Sci. Lett.* **302** L263

Sanders H. E., Gardner P., King D. A. and Morris M. A. (1994) *Surf. Sci.* **304** 159

Sanderson R. T. (1960) in *Chemical Periodicity* (Rheinhold: New York)

Sanderson R. T. (1964) *Inorg. Chem.* **3** 925

Sanz J. M., Gonzalez-Elipe A. R., Fernandez A., Leinen D., Galan L., Stampfl A. and Bradshaw A. M. (1994) *Surf. Sci.* **307–309** 848

Sauer J. (1989) *Chem. Rev.* **89** 199

Sawada S. and Nakamura K. (1979) *J. Phys. C: Solid State Phys.* **12** 1183

Sayle T. X. T., Parker S. C. and Catlow C. R. A. (1994) *Surf. Sci.* **316** 329

Scamehorn C. A., Hess A. C. and McCarthy M. I. (1993) *J. Chem. Phys.* **99** 2786

Scamehorn C. A., Harrison N. M. and McCarthy M. I. (1994) *J. Chem. Phys.* **101** 1547

Schlüter M. (1978) *Phys. Rev.* **B17** 5044

Schlüter M. (1982) *Thin Solid Films* **93** 3

Schmalzried H. and Backhaus-Ricoult M. (1993) *Prog. Solid State Chem.* **22** 1

Schönberger U., Andersen O. K. and Methfessel M. (1992) *Acta Metall. Mater.* **40** S1

Schottky W. (1939) *Z. Phys.* **113** 367

Schultz J. and Gent A. N. (1973) *J. Chim. Phys.* **70** 708

Schulz K. H. and Cox D. F. (1991) *Phys. Rev.* **B43** 1610

Seiyama T. (1978) in *Metal Oxides and their Catalytic Actions* (Kodansha Scientific: Tokyo)

Selverian J. H., Ohuchi F. S., Bortz M. and Notis M. R. (1991) *J. Materials Sci.* **26** 6300

Shannon R. D. (1976) *Acta Cryst.* **A32** 751

Sharma R. R. and Stoneham A. M. (1976) *J. Chem. Soc. Faraday Trans. II* **72** 913

Shibata K., Kiyoura T., Kitagawa J., Sumiyoshi T. and Tanabe K. (1973) *Bull. Chem. Soc. Jpn.* **46** 2985

Shido T., Asakura K. and Iwasawa Y. (1989) *J. Chem. Soc. Faraday Trans. I* **85** 441

Shih A., Hor C., Mueller D., Marrian C. R. K., Elam W. T., Wolf P., Kirkland J. P. and Neiser R. A. (1988) *J. Vac. Sci. Technol.* **A6** 1058

Shirane G. and Yamada Y. (1969) *Phys. Rev.* **177** 858

Silvi B., Fourati N., Nada R. and Catlow C. R. A. (1991) *J. Phys. Chem. Solids* **52** 1005

Skinner A. J. and LaFemina J. P. (1992) *Phys. Rev.* **B45** 3557

Slater R. R. (1970) *Surf. Sci.* **23** 403

Smith D. J., Bursill L. A. and Jefferson D. A. (1986) *Surf. Sci.* **175** 673

Solomon E. I., Jones P. M. and May J. A. (1993) *Chem. Rev.* **93** 2623

Sousa C., Illas F., Bo C. and Poblet J. M. (1993a) *Chem. Phys. Lett.* **215** 97

Sousa C., Illas F. and Pacchioni G. (1993b) *J. Chem. Phys.* **99** 6818

Spanjaard D. and Desjonquères M. C. (1984) *Phys. Rev.* **B30** 4822

Spicer W. E., Lindau I., Skeath P. and Yu C. Y. (1980) *J. Vac. Sci. Technol.* **17** 1019

Spicer W. E., Kendelewicz T., Newman N., Chin K. K. and Lindau I. (1986) *Surf. Sci.* **168** 240

Stefanovich E. V., Shluger A. L. and Catlow C. R. A. (1994) *Phys. Rev.* **B49** 11560

Stein J. and Prutzman L. C. (1988) *J. Appl. Polymer Sci.* **36** 511

Steinruck H. P., Pesty F. and Madey T. E. (1995) *Phys. Rev.* **B51** 2427

Stoneham A. M. (1983) *Appl. Surf. Sci.* **14** 249

Stoneham A. M. and Harding J.H. (1986) *Ann. Rev. Phys. Chem.* **37** 53

Stoneham A. M. and Tasker P. W. (1987) *J. Physique (France)* **49** supplément **C5** 99

Stoneham A. M. and Tasker P. W. (1988) in *Surface and Near Surface Chemistry of Oxide Materials*, Materials Science Monographs **47** (Eds.: Nowotny J. and Dufour L. C.) (Elsevier: Amsterdam) p. 1

Sugano S. and Shulman R. G. (1963) *Phys. Rev.* **130** 517

Sulimov V. B., Pisani C. and Cora F. (1994) *Solid State Comm.* **90** 511

Tait R. H. and Kasowski R. V. (1979) *Phys. Rev.* **B20** 5178

Takada A., Catlow C. R. A., Lin J. S., Price G. D., Lee M. H., Milman V. and Payne M. C. (1995) *Phys. Rev.* **B51** 1447

Tanabe K. (1970) *Solid Acids and Bases* (Academic Press: New York, London)

Tanabe K. (1978) in *Metal Oxides and Complex Oxides* (Eds.: Tanabe K., Seiyama T. and Hueki K.) (Kodansha Scientific: Tokyo) p. 407

Tanabe K. (1981) in *Catalysis, Science and Technology* (Eds.: Boudert M. and Anderson J. R.) (Springer Verlag: Berlin, Heidelberg, New York), chapter 5

Tanabe K. and Fukuda Y. (1974) *React. Kinet. Catal. Lett.* **1** 21

Tanabe K., Misono M., Ono Y. and Hattori H. (1989) in *New Solid Acids and Bases, their Catalytic Properties* (Eds.: Belmon B., Yates J. T.) (Elsevier: Amsterdam) p. 51

Tanaka H., Matsumoto T., Kawai T. and Kawai S. (1994) *Surf. Sci.* **318** 29

Tanaka K. and Ozaki A. (1967) *J. Catal.* **8** 1

Tasker P. W. (1979a) *J. Phys. C: Solid State Phys.* **12** 4977

Tasker P. W. (1979b) *Surf. Sci.* **87** 315

Tasker P. W. (1984) *Adv. Ceram.* **10** 176

Tasker P. W. and Duffy D. M. (1984) *Surf. Sci.* **137** 91

Tasker P. W. and Stoneham A. M. (1987) *J. Chim. Phys.* **84** 149

Taverner A. E., Hollamby P. C., Alridge P. S., Egdell R. G. and Mackrodt W. C. (1993) *Surf. Sci.* **287–288** 653

Tersoff J. (1985) *J. Vac. Sci. Technol.* **B3** 1157

Texter J., Klier K. and Zettlemoyer A. C. (1978) *Prog. Surf. Memb. Sci.* **12** 327

Thibado P. M., Röhrer G. S. and Bonnell D. A. (1994) *Surf. Sci.* **318** 379

Thiel P. A. and Madey T. E. (1987) *Surf. Sci. Rep.* **7** 211

Thiry P., Chandesris D., Lecante J., Guillot C., Pinchaux R. and Petroff Y. (1979) *Phys. Rev. Lett.* **43** 82

Thiry P. A., Liehr M., Pireaux J. J. and Caudano R. (1984) *Phys. Rev.* **B29** 4824

Thiry P. A., Liehr M., Pireaux J. J., Sporken R., Caudano R., Vigneron J. P. and Lucas A. A. (1985) *J. Vac. Sci. Technol.* **B3** 1118

Timsit R. S., Waddington W. G., Humphreys C. J. and Hutchinsons J. L. (1985) *Ultramicroscopy* **18** 387

Tjeng L. H., Vos A. R. and Sawatzky G. A. (1990) *Surf. Sci.* **235** 269

Tong S. Y., Xu G. and Mei W. N. (1984) *Phys. Rev. Lett.* **52** 1963

Tosi P. (1964) *Solid State Phys.* **16** 1

Toussaint G., Selme M. O. and Pecheur P. (1987) *Phys. Rev.* **B36** 6135

Towler M. D., Allan N. L., Harrison N. M., Saunders V. R., Mackrodt W. C. and Apra E. (1994) *Phys. Rev.* **B50** 5041

Trampert A., Ernst F., Flynn C. P., Fishmeister H. F. and Rühle M. (1992) *Acta Metall. Mater.* **40** S227

Trumble K. P. and Rühle M. (1991) *Acta Metall. Mater.* **39** 1915

Tsuda N. and Fujimori A. (1981) *J. Catal.* **69** 410

Tsukada M. and Hoshino T. (1982) *J. Phys. Soc. Jpn.* **51** 2562

Tsukada M., Miyazaki E. and Adachi H. (1981) *J. Phys. Soc. Jpn.* **50** 3032

Tsukada M., Adachi H. and Satoko C. (1983) *Prog. Surf. Sci.* **14** 113

Turchi P., Ducastelle F. and Treglia G. (1982) *J. Phys. C: Solid State Phys.* **15** 2891

Ugliengo P., Saunders V. and Garrone E. (1990) *J. Phys. Chem.* **94** 2260

Underhill P. R. and Gallon T. E. (1982) *Solid State Comm.* **43** 9

Urano T., Kanaji T. and Kaburagi M. (1983) *Surf. Sci.* **134** 109

Usanovich M. I. (1939) *J. Gen. Chem. USSR* **9** 182

Varma S., Chottiner G. S. and Arbad M. (1992) *J. Vac. Sci. Technol.* **A10** 2857

Varma S., Chen X., Zhang J., Davoli I., Saldin D. K. and Tonner B. P. (1994) *Surf. Sci.* **314** 145

Ventrice Jr. C. A., Bertrams T., Hannemann H., Brodde A. and Neddermeyer H. (1994) *Phys. Rev.* **B49** 5773

Vermeersch M., Sporken R., Lambin P. and Caudano R. (1990) *Surf. Sci.* **235** 5

Vermeersch M., Malengreau F., Sporken R. and Caudano R. (1995) *Surf. Sci.* **323** 175

Verwey E. J. W. (1946) *Recl. Trav. Chim.* **65** 521

Vinek H., Noller H., Ebel M. and Schwarz K. (1977) *J. Chem. Soc. Faraday Trans. I.* **73** 734

Vogtenhuber D., Podloucky R., Neckel A., Steinmann S. G. and Freeman A. J. (1994) *Phys. Rev.* **B49** 2099

Wang C. R. and Xu Y. S. (1989) *Surf. Sci.* **219** L537

Wang L., Liu J. and Cowley J. M. (1994) *Surf. Sci.* **302** 141

Warren O. L. and Thiel P. A. (1994) *J. Chem. Phys.* **100** 659

Weiss W., Barbieri A. and Van Hove M. A. (1993) *Phys. Rev. Lett.* **71** 1848

Welton-Cook M. R. and Berndt W. (1982) *J. Phys. C: Solid State Phys.* **15** 5691

de Wette F. W., Kress W. and Schroeder U. (1985) *Phys. Rev.* **B32** 4143

Wikborg E. and Inglesfield J. E. (1977) *Physica Scripta* **15** 37

Winterbottom W. L. (1967) *Acta Metall.* **15** 303

Wolf D. (1992) *Phys. Rev. Lett.* **68** 3315

Wolfram T., Krant E. A. and Morin F. J. (1973) *Phys. Rev.* **B4** 1677

Wu M. C., Estrada C. A. and Goodman D. W. (1991a) *Phys. Rev. Lett.* **67** 2910

Wu M. C., Corneille J. S., Estrada C. A., He J. W. and Goodman D. W. (1991b) *Chem. Phys. Lett.* **182** 472

Wulser K. W. and Langell M. A. (1994) *Surf. Sci.* **314** 385

Wulser K. W., Hearty B. P. and Langell M. A. (1992) *Phys. Rev.* **B46** 9724

Xu X., He J. W. and Goodman D. W. (1993) *Surf. Sci.* **284** 103

Xu Y. N. and Ching W. Y. (1991) *Phys. Rev.* **B43** 4461

Xu Y. N., Ching W. Y. and French R. H. (1990) *Ferroelectrics* **111** 23

Yamanaka T. Tanabe K. (1975) *J. Phys. Chem.* **79** 2409

Yamanaka T. Tanabe K. (1976) *J. Phys. Chem.* **80** 1723

Yang W. and Parr R. G. (1985) *Proc. Natl. Acad. Sci. USA* **82** 6723

Yin M. T. and Cohen M. L. (1982) *Phys. Rev.* **B26** 5668

Zaanen J. and Sawatzky G. A. (1990) *J. Solid State Chem.* **88** 8

Zaanen J., Sawatzky G. A. and Allen J. W. (1985) *Phys. Rev. Lett.* **55** 418

Zhang Z. and Henrich V. E. (1992) *Surf. Sci.* **277** 263

Zhang Z. and Henrich V. E. (1994) *Surf. Sci.* **321** 133

Zhang Z., Jeng S. P. and Henrich V. E. (1991) *Phys. Rev.* **B43** 12004

Zhong Q. and Ohuchi F. S. (1990) *J. Vac. Sci. Technol.* **A8** 2107

Zhong Q., Vohs J. M. and Bonnell D. A. (1992) *Surf. Sci.* **274** 35

Zhou J. B., Lu H. C., Gustafsson T. and Haberle P. (1994) *Surf. Sci.* **302** 350

Ziman J. M. (1964) *Principles of the Theory of Solids* (Cambridge University Press: Cambridge)

Ziemann P. J. and Castelman Jr A. W. (1991) *J. Chem. Phys.* **94** 718

Zschack P., Cohen J. B. and Chung Y. W. (1992) *Surf. Sci.* **262** 395

Index